AF541332

PAKISTAN, ISI AND HUNT FOR TERROR

PAKISTAN, ISI AND HUNT FOR TERROR

Col. Indrajeet Singh(Retd.)

GAURAV BOOK CENTRE PVT LTD
DELHI

Publisher
GAURAV BOOK CENTRE PVT LTD
4832/24,Prahlad Lane,S-207 Ansari
Road, Daryaganj, Delhi-110002
Ph.: 43570976, 23278261
Email: gauravbookcentre@gmail.com

Edition: 2015

ISBN: 978-93-83316-08-3

Laser Typesetting
JEE-VEE Graphics, Delhi

Price: 1195/-

Printed
Vikas Computers, Delhi

Preface

The ISI was founded in 1948 to facilitate intelligence gathering and sharing between the three main sections of the armed forces: the army, navy and air force. There are other military and civilian intelligence agencies too, but the ISI is undoubtedly the most powerful and the most politicised among them.

The heavily guarded, high-walled concrete structure right in the heart of Islamabad has no signboard. But everyone in the capital knows that the sprawling compound surrounded by barbed wire houses the Directorate for Inter-Services Intelligence (ISI). The spy agency has been the country's big brother. It is powerful, ubiquitous and has functioned with so much autonomy from the central government that it has almost become a state within a state. It is not only responsible for intelligence gathering, but also acts as a determinant of Pakistan's foreign policy and a vehicle for its implementation. For every civilian and military government, control of the ISI was seen as crucial to maintaining a firm grip on power. The agency had been so powerful for so long that it played by its own rules. Its various heads had contrasting profiles, but emerged among the most powerful figures in the country's establishment. For years they ran semi-independent operations in Afghanistan and Kashmir and helped to form and topple civilian governments.

Since Islamabad joined Washington as an ally in the post-9/11 "war on terror," analysts have accused Pakistan's security and intelligence services of playing a "double game," tolerating if not outright aiding militant groups killing NATO troops in Afghanistan. Pakistan denies these charges.

—Editor

Contents

1

The Importance of Pakistan

ETHNIC GROUPS IN PAKISTAN

Islamabad does not need to worry; regardless of the bin Laden episode, Washington will not walk out of their so-called alliance against terror. The United Sates needs Pakistan for its operation in Afghanistan and does not want China to take its place in the region. For these two reasons, the Obama Administration will close its eyes to many non-senses of the ISI and the Pakistani Army.

A year before 9/11 attack on the twin towers, the US Under-Secretary of State Thomas Pickering met with some Pakistani officials in New York. One of the issues discussed was bin Laden, Washington was already hunting the Saudi terrorist: "Pickering opened the meeting by expressing disappointment that Pakistan, whom he called a good friend of the US, was not taking steps to help with OBL," says a secret cable addressed to the US Secretary of State.

The United Sates needs Pakistan for its operation in Afghanistan and does not want China to take its place in the region. The same tune continued till the Abbottabad raid: Pakistan never helped (on the contrary) to catch any terrorist on its soil, but the US and Pakistan remain 'friends'.

In fact 'friend' is not the correct word; Pakistan is more of a vassal State which gets yearly lollipops to remain quiet. A historian wrote: "an Empire has no allies; it has only vassal nations who in principle are sovereign, but let themselves be reduced to the sad condition of satellites." Is not America the last colonial empire? For the past fifty years, the position of the United States vis-à-vis Pakistan has to be seen in this perspective. It is the tragedy of Pakistan, built on a falsehood (The Two Nation Theory), that it requires a big brother to (economically) stand on its legs. Washington may not be happy with its vassal for having given refuge to Enemy no 1, right in the midst of a military cantonment, but the fact remains that successive governments in Islamabad have protected bin Laden (and so many other terrorists and jihadists), while Washington has continued to financially keep Islamabad afloat.

In fact friend is not the correct word; Pakistan is more of a vassal State which gets yearly lollipops to remain quiet. Colonel Sir Francis Younghusband who became famous after his military expedition in Tibet in 1904, once wrote: "We, who have dealt with Asiatics, can appreciate so well [the following tactic]: taking the opportunity, striking while the iron is hot, not letting the chance go by, knowing our mind, knowing what we want, and acting decisively when the exact occasion arises." The occasion arose, Obama struck; but the relation between the two 'friends' remains unchanged.

In July 1946, the British Indian Chiefs of Staff Committee prepared an interesting report "on the importance of India as a military base". C. Dasgupta, the retired diplomat who wrote *War and Diplomacy in Kashmir* pointed out: " most significant feature, however, was the re-assessment of India's geostrategic value. This reflected a notable shift of emphasis from the naval to the air factor."

Obama struck; but the relation between the two friends remains unchanged. The British Government, would it decide to leave the subcontinent, needed 'air bases' to be able to keep a tab on Afghanistan, Central Asia, Soviet Union and Eastern Turkestan. Quoting from the British report, Dasgupta continued: "It was also essential that the Soviet Union be denied air bases in India. If India was dominated by Russia with powerful air forces it is likely that we should have to abandon our command of the Persian Gulf and the Northern Indian Ocean routes, making it impossible to ensure the uninterrupted flow of vitally important oil supplies."

Another report prepared in London came to the same conclusions: "India's territory provides important bases for offensive air action and for the support of our forces in the Indian Ocean and neighbouring areas". Having decided to partition the sub-continent, London understood that Pakistan was the best bet strategically. In October 1946, the India Office advised the Chiefs of Staff Committee that "if India were to split up into two or more parts, the Muslim areas would probably be anxious to remain in the Commonwealth if, in such circumstances, we were willing to have them".

"India's territory provides important bases for offensive air action and for the support of our forces in the Indian Ocean and neighbouring areas". For the British (and later the Americans), 'Hindus' were unreliable allies, having strange ideas such as non-violence or non-alignment as the main features of their future foreign policy.

The Partition was therefore a good solution for London; they could keep air bases in Pakistan and at the same time walk out of the sub-continent without too much damage. The Americans followed the footsteps of the British in the early 1950's. They had an even better idea: an independent Kashmir.

As early as January 1948, Sheikh Abdullah was ready to offer the Maharaja's State on a platter to the Americans, but it took some five years for Washington to realize how serious the proposal was. When Adlai Stevenson came to Srinagar in May 1953, he discussed with Abdullah the creation of an independent Sheikhdom of Kashmir state. It suited perfectly the American interests as it could thus check the advances of the Chinese in Sinkiang and the Russians in Afghanistan.

Just a look at a map of the Pakistan shows the extraordinary strategic importance of the country. Kashmir controlled the roads to Afghanistan, to the central Asian republics of Soviet Union, to China and to Tibet. A few weeks after the meeting, Abdullah asserted: "It is not necessary that our State should become an appendage of either India or Pakistan." Only after he was arrested in August 1953, did the Americans drop their idea to have a base in Kashmir. Hardly a year later, Washington's sight turned towards Pakistan; it was the beginning of the famous 'friendship'. Just a look at a map of the Pakistan shows the extraordinary strategic importance of the country. The British were no fools! They knew it. Already way back in 1877, a Political Agent was posted in Gilgit to keep an eye on the region. Since they followed the footsteps of the British, America had been looking for an obedient ally to keep a foot in the region. Today another player has entered the Great Game, Beijing. To come out their vassalage, Islamabad is now cleverly playing the Chinese card. A report from Pakistan suggests that the Chinese are interested in studying the remains of the US top-secret stealth helicopter abandoned during the Abbottabad raid. A Pakistani official even said, "We might let them [the Chinese] take a look."

It is a way to convey the message: the Chinese are waiting at our door, "don't mess with us, our 'all-weather friend' can replace you". "the Chinese are waiting at our door, "don't mess with us, our "all-weather friend can replace you".

On May 7, 2011, an interesting article was published by *China Review News*, "The More Anti-terrorism, The More Terror: the Post-bin Laden Era Tests US Strategy". The author predicts that China will be the next target of US anti-terrorism 'war'. The argument is that before 9/11, President Bush viewed China as the main strategic rival and exerted diverse forms of pressure and then when "Osama bin Laden gave a vicious blow to the US, George W. Bush became a different person and turned to cooperate with China". The conclusion is "9/11 changed the foundations of Sino-US relations and gave China a 10-year golden opportunity to regain strength." Now a new era of uncertainties has begun, however Washington has no choice but to remain 'friends' with Pakistan, at least till Jinnah's State implodes under its own contradictions.

PAKISTAN: WORSE THAN WE KNEW

A pro-Taliban rally in Quetta, the capital of Pakistan's Balochistan province, circa 2002. During the Afghan elections in early April I was traveling in Central Asia, mainly in Kyrgyzstan. I wanted to inquire into the fears of the governments there as a result of the US withdrawal from Afghanistan. What did they think of the growth of Taliban and Islamic extremism in Afghanistan and Pakistan? Officials in each country cited two threats. First, the internal radicalizing of their young people by increasing numbers of preachers or proselytizing groups arriving from Pakistan, Bangladesh, and the Middle East. The second, more dangerous threat is external: they believe that extremist groups based in Pakistan and Afghanistan are trying to infiltrate Central Asia in order to launch terrorist attacks.

Islamic extremism is infecting the entire region and this will ultimately become the legacy of the US occupation of Afghanistan, as the so-called jihad by the Taliban against the US comes to an end. Iran, a Shia state, fears that the Sunni extremist groups that have installed themselves in Pakistan's Balochistan province on the Iranian border will step up their attacks inside Iran. In February Iran threatened to send troops into Balochistan unless Pakistan helped free five Iranian border guards who had been kidnapped by militants. (The Pakistanis freed four of the guards; one was killed.) Chinese officials say they are particularly concerned about terrorist groups coming out of Pakistan and Afghanistan that are undermining Chinese security. Although China is Pakistan's closest ally, its officials have made it clear that they are closely monitoring the Uighur Muslims from Xinjiang province, who are training in Pakistan, fighting in Afghanistan, and have carried out several terrorist attacks in Xinjiang.

Terrorist assaults from Pakistan into Indian Kashmir have declined sharply since 2003, but India has a perennial fear that Islamic militant groups based in Pakistan's Punjab province may mount attacks in India. Many Punjabi fighters have joined the Taliban forces based in Afghanistan and in Pakistan, and they have attacked Indian targets in Afghanistan. India is also wary of another terrorist attack resembling the one that took place in Mumbai in 2008.

For forty years Pakistan has been backing Islamic extremist groups as part of its expansionist foreign policy in Afghanistan and Central Asia and its efforts to maintain equilibrium with India, its much larger enemy. Now Pakistan is undergoing the worst terrorist backlash in the entire region. Some 50,000 people have died in three separate and continuing insurgencies: one by the Taliban in the northwest, the other in Balochistan by Baloch separatists, and the third in Karachi by several ethnic groups. That sectarian war, involving suicide bombers, massacres, and kidnappings, has gripped the country for a decade.

Some five thousand Pakistani soldiers and policemen have been killed and some twenty thousand wounded, both as targets of terrorist attacks and during offensives against them. The economy has sharply declined, and there are widespread electricity shortages. The political elite is divided and at odds with the military over how to deal with terrorism, while many in the middle class are leaving the country.

Two years ago all the states in the region would have publicly or privately accused Pakistan's military and Interservices Intelligence (ISI) of supporting, protecting, or at least tolerating almost every terrorist group based in Pakistan. The ISI had links with all of them and often collaborated with them. Recently those relations have changed. Governments in the region now accept that Pakistan is in some ways trying to fight terrorism on its soil. But those governments are also concerned that the Pakistani military and political elite have lost control of large parts of the country and cannot maintain law and order. The US and Western countries fear that Pakistan's nuclear weapons arsenal is vulnerable and that terrorists in Pakistan may be planning an attack comparable to that of September 11.

There is still no overall political or military strategy to combat Islamic extremism. The Pakistani army tries to suppress some terrorist groups but not, for example, those that target India. Such a selective strategy cannot be maintained indefinitely and poses enormous risks to the entire world.

Since the mid-1970s the ISI has supported extremist Islamic groups in Afghanistan including the Taliban, but that policy may now be changing. Contrary to many predictions, the situation in Afghanistan may be taking a turn for the better. Despite the threat of Taliban reprisals, seven million Afghans turned out on April 5 to vote in the first presidential election in which President Hamid Karzai was not a candidate. This was also the first genuine attempt in Afghan history to transfer power democratically. A remarkable 58 percent of the 12 million eligible voters turned out—35 percent of them women. Although the Taliban did not make a show of force to stop the vote, relatively few people voted in many Taliban-controlled areas in the south and east. Preliminary results released on April 26 show the Tajik leader Abdullah Abdullah in the lead with 45 percent of the vote and his Pashtun rival Ashraf Ghani trailing with 32 percent. Over three thousand cases of fraud still have to be investigated before the count is final.

Since neither candidate had a majority of 50 percent, there will be a runoff election between the two by the end of May. A new government will not be in place before July, which means that a security agreement with the US, which all the candidates have agreed to, will be delayed. The US and NATO want a military force of some ten thousand to stay in the country in order to train the Afghan army and gather intelligence. Such an agreement will be necessary if the US Congress and Europe are to be persuaded to keep the Kabul government financially afloat. Afghanistan needs a minimum of $7 billion a year to pay for its budget and army. In January the US Congress cut by half the $2 billion earmarked for US aid to Afghanistan.

To bring the civil war to an end the new president will try to open talks with the Pashtun Taliban in Afghanistan. Pakistan is now also keen on such talks because two thirds of Pashtuns live in Pakistan, including members of the Taliban, and there has been talk by Islamists of carving out a separate Pashtun state. Will the Pakistan military put pressure on the Afghan Taliban leaders who live in Pakistan to talk to the new government in Kabul while Pakistan deals with its own Pashtun problem? A lot will depend on whether a much weakened Pakistan still has the power to force the Afghan Taliban to engage in negotiations.

All the recent books I have seen on the Afghan wars have recounted how the Pakistani military backed the Taliban when they first emerged in 1993, but lost its influence by 2000. Then, after a brief respite following September 11, 2001, Pakistan's military helped to resurrect the Taliban resistance to fight the Americans. My own three books on Afghanistan describe the actions of the Pakistani military as one factor in keeping the civil war going and contributing to the American failure to win decisively in Afghanistan.

Now in *The Wrong Enemy: America in Afghanistan, 2001–2014*, Carlotta Gall, the*New York Times* reporter in Afghanistan and Pakistan for more than a decade, has gone one step further. She places the entire onus of the West's failure in Afghanistan and the Taliban's successes on the Pakistani military and the Taliban groups associated with it. Her book has aroused considerable controversy, not least in Pakistan. Its thesis is quite simple:

The [Afghan] war has been a tragedy costing untold thousands of lives and lasting far too long. The Afghans were never advocates of terrorism yet they bore the brunt of the punishment for 9/11. Pakistan, supposedly an ally, has proved to be perfidious, driving the violence in Afghanistan for its own cynical, hegemonic reasons. Pakistan's generals and mullahs have done great harm to their own people as well as their Afghan neighbors and NATO allies. Pakistan, not Afghanistan, has been the true enemy.

Dogged, curious, insistent on uncovering hidden facts, Gall's reporting over the years has been a nightmare for the American, Pakistani, and other foreign powers involved in Afghanistan, while it has been welcomed by many Afghans. She quickly emerged as the leading Western reporter living in Kabul. She made her reputation by reporting on the terrible loss of innocent Afghan lives as American aircraft continued to bomb the Pashtun areas in southern Afghanistan even after the war of 2001 had ended. The bombing of civilians was said to be accidental, supposedly based on faulty intelligence; but it continued for years and helped the Taliban turn the population against the Americans.

Before human rights groups or police arrived in remote, bombed villages, Gall was often there first. Thus in July 2002, she writes of driving "for three days over dusty and rutted roads" to reach a village in Uruzgan province that had been bombed during a wedding. Fifty-four wedding guests were killed, including thirteen children from one household, and over one hundred people were wounded. The survivors of this massacre "were collecting body parts in a bucket"—Gall's quote of the provincial governor that haunted reporters and other observers in Kabul. She continued:

> Sahib Jan, a twenty-five-year-old neighbor, was one of the first to reach the groom's house after the bombardment. Bodies were lying all over the two courtyards and in the adjoining orchard, some of them in pieces. Human flesh hung in the trees. A woman's torso was lodged in an almond sapling.... Bodies lay in the dust and rubble of the rooms below.

Some of those killed were friends of President Karzai and these bombings infuriated him and caused his relations with the US to deteriorate. As late as 2009 Gall was still covering such disasters, as when US planes bombed the village of Granai, killing 147 people—"the worst single incident of civilian casualties of the war."

Carlotta Gall was, in effect, a one-woman human rights agency. She spent much time and effort exposing the torture and killing of Afghans taken prisoner by the Americans. This was a highly sensitive issue—the American victors did not expect American media to expose their wrongdoings. But Gall went ahead. She told the heartbreaking story of Dilawar, a naive taxi driver who was wrongly arrested in Khost in eastern Afghanistan, incarcerated in an isolation ward at the US airbase at Bagram, and then beaten to death by his American jailors. She spent many weeks tracking down Dilawar's family and obtained the death certificate issued by the US Army: I gasped as I read it. I had been looking to learn more about the Afghans being detained. I had not expected to find a homicide committed by American soldiers.

Nobody was ever charged and the same US team of interrogators was deployed to Abu Ghraib in Iraq—the other site of grisly US treatment of prisoners. Gall's modesty does not allow her to mention that it was this story that led to the making of the 2007 Oscar-winning documentary *Taxi to the Dark Side*.

All her skills were put to the test when she reported on the death of Osama bin Laden in Abbottabad and tried to discover whether senior Pakistanis had been hiding him all along. Methodically adding one fact to another, she concludes not only that some were, which is convincing, but that all the top officials in the military and the ISI knew of his whereabouts, although the evidence she offers for such widespread knowledge is not wholly plausible; and her assertion that there was a specific "bin Laden desk" at the ISI appears, from my own inquiries, to be flimsy.

For many Pakistanis the main failure of the government is that nobody has ever been punished or held responsible either for hiding bin Laden or not discovering him earlier. Gall surmises that the ISI had let it be known that bin Laden's hideaway was an ISI safe house. That is why nobody ever knocked on the door—a reasonable assumption.

However, the fiercest opposition to her views comes from American officials themselves. They insist, as they are obliged to do, that none of the top Pakistani leaders knew of bin Laden's whereabouts. Gall's conclusion that the Obama administration deliberately kept the ISI's role in harbouring bin Laden secret in order to save the US–Pakistan relationship is difficult to accept for two reasons. The first is simply the propensity of officials in Washington to leak to journalists. The second is that US–Pakistani relations would collapse a few months after the killing of bin Laden over different issues, notably Pakistan's support of the Taliban. The US therefore would not have been so concerned to protect its relations with Pakistan.

Most states today, including the US and NATO countries, believe that the Pakistani military is no longer in control of the Taliban in Afghanistan or capable of putting decisive pressure on them. The army leaders have too much of a problem at home with their own Pakistani Taliban. Their ability to persuade the Afghan Taliban to make peace with Kabul is very limited. Moreover, the Pakistani military has shown no willingness to kick the Afghan Taliban out of Pakistan and back to Afghanistan. The civilian government is trying to negotiate with the Pakistani Taliban but the military is against such talks and would rather use force, a major division in policymaking in Islamabad. There are enormous risks involved, such as the two Talibans merging to fight the

Pakistani army. The Pakistani military belatedly understands that a Taliban conquest in Afghanistan would eventually ensure that Pakistan would find itself with a Taliban government in Islamabad. As Gall recounts, the Pakistani army has spent years propping up the Afghan Taliban, training their fighters, allowing them to import arms and money from the Arabian Gulf and to recruit among Pakistani youth. As Gall shows, the army even decided which tactics the Afghan Taliban should use. The army is now desperate to find a political solution that would send the Afghan Taliban home.

Many army and police officers find themselves confused as they are ordered to protect some Taliban and other extremists and kill others. Pakistani officials are supposed to be loyal allies of the US and they take its money but they also are encouraged by powerful Pakistanis to promote anti-Americanism in society and the army. There has been no adequate explanation for these dual-track policies, which have ravaged state and society and undermined the army internally. Moreover the army is still not prepared to give up its militant stand against India.

Gall writes that Pakistani soldiers "were fighting, and dying, in campaigns against Islamist militants, apparently at the request of America, but at the same time they were being fed a constant flow of anti-American and pro-Taliban propaganda." Unfortunately she does not acknowledge that there have been shifts in the military's thinking and that it faces the more open kind of confusion over its strategy and its loyalties that I have described. Her book starts and ends on the same note even though thirteen years have elapsed.

Afghans have observed that the ISI has not interfered in the Afghan elections. Contrary to its policy since the 1970s, it has avoided favouring Pashtun candidates. It has also tried to improve relations with the former anti- Taliban Northern Alliance (NA) warlords it once opposed by meeting with the leaders of Tajik, Hazara, and Uzbek groups that were the major components of the alliance. Consequently all the Afghan presidential candidates have softened their comments on Pakistan, avoiding the harsh rhetoric of Hamid Karzai.

Yet for the reasons described by Gall, the Pakistani military still does not comprehend how deeply Pakistan is hated by most Afghans. Even today the worst atrocities and suicide bombings causing civilian deaths are often blamed on the Taliban elements "trained by Pakistanis." Hatred for Pakistan is possibly even stronger among the Afghan Pashtuns who have been Pakistan's traditional allies. The Pakistani army must undergo deep self-examination and show considerable humility in dealing with the Afghans if it is to genuinely create an opportunity for peace.

However there are large gaps in Gall's analysis that cannot be ignored. Pakistan was not the only cause of the failure to control the Afghan Taliban; the failure in Afghanistan has been an American failure as well. The lack of a US political strategy stretched over four administrations. Two Presidents—Bush and Obama—were unable to make up their minds about what to do in Afghanistan or how many troops should carry out which tasks. The overwhelming militarization of US decision-making and the hubris of American generals undermined diplomacy and nation-building; the US failed to curb open production of opium and other drugs. There was constant infighting between the White House, Defense, and State Departments over policy. There was also widespread corruption and waste both in the private contracting system used by the US military and in some of the operations of the US Agency for International Development. The list of such American failures is indeed long, and assigning responsibility for the losses in Afghanistan will occupy US historians for decades.

Gall's second omission is not to recognize the negative effects caused by the neighbouring countries, apart from Pakistan, and their constant interference in Afghanistan. She ignores the Afghan civil war after 1989 when all the Afghan warlords had international backers. She fails to mention that Pakistan and Saudi Arabia backed the Taliban while Russia, Iran, India, Turkey, and the Central Asian republics supported the Northern Alliance.

More recently Iran has given sanctuary to the Taliban and al-Qaeda, India is funding the Baloch separatist insurgency in Pakistan, and Afghanistan has provided a refuge to the leader of the Pakistani Taliban. The US presence has failed to provide protection for people in the region. Most Afghans will tell you today that what they fear most about the Americans leaving is that intervention from all the country's neighbors will start again. Gall doesn't blame neighbors other than Pakistan.

Why did Pakistan adopt policies of intervention in Afghanistan, especially after September 11, when it had essentially lost the game in Afghanistan? There has been a disastrous logic to the military's policies—which more thoughtful Pakistanis have always resisted.

Here some history is useful. The Pakistan military has used militant political groups as an arm of its foreign policy in India and Afghanistan since the 1970s. This was allowed by the West as part of the cold war. During the 1980s the CIA funded the Afghan Mujahideen and Islamic extremists from forty countries when they were fighting the Soviet occupation of Afghanistan. It was not until September 11 that Pakistan's use of Islamic extremists as a tool of its foreign policy became unacceptable.

After September 11 General Pervez Musharraf and the military regime believed that they could, for a time, appear to meet US demands by capturing al-Qaeda leaders while avoiding harm to the Afghan Taliban. Musharraf was always treated as a messiah by the Bush administration; but a year after September 11 well-informed Pakistanis knew that Musharraf had started playing a double game with the Americans by covertly supporting a Taliban resurgence.

What was the Pakistan military's logic in doing so? After the war to oust the Taliban was over in 2001 the military faced the defeat of its Taliban allies and had to suffer the Northern Alliance and its backers—including India and Iran—as victors in Kabul. Musharraf felt he had to preserve some self-respect; and Bush appeared to acknowledge this when he allowed ISI agents to be airlifted out of Kunduz before the city fell to the Northern Alliance and its backers—a series of events well described by Gall. Bush had also promised Musharraf that the NA would not enter Kabul before a neutral Afghan body under the UN took over the city. But as the Taliban fled, the NA walked into Kabul without a fight and took over the government.

The Pakistani military was further angered at Bonn in December 2001, when the new Afghan government was unveiled and all the provincial security ministries were handed over to the Northern Alliance, with Pashtun representation at a minimum. This was the usual outcome by which the spoils of war went to the victors, but for Pakistan's generals it was further humiliation that bred resentment and a desire for revenge.

The military was equally perplexed about why the US did not commit more ground troops to hunt down al-Qaeda instead of leaving that task to Northern Alliance warlords. The military was convinced that the Americans would soon abandon Afghanistan for the war in Iraq and leave the NA, backed by India, in charge in Kabul.

Bush's refusal to commit even one thousand US troops to the mountains of Tora Bora where bin Laden was trapped sent a powerful message to Pakistan. By 2003 US forces in Afghanistan still amounted to only 11,500 men—insufficient to hold the country. Five years later in 2008 there were only 35,000 US troops in Afghanistan, compared to five times that number in Iraq.

The Pakistani military's insecurity about American intentions and the growing power of the NA, India, and Iran led to its fateful decision to rearm the Taliban. It believed that the Taliban would provide a form of protection for the Pakistani military against its enemies. Instead the revamped Afghan Taliban helped create the Pakistani Taliban and the worst blowback of terrorism in Pakistan's history. It is the Taliban's terrorism within Pakistan rather than US pressure that altered the military's position from backing the Afghan Taliban to its now seeking a peaceful Afghanistan.

Gall's account of the rise of the Taliban is also open to question. She writes that three commanders in Kandahar and Kabul—two of them drug smugglers and one of them a landlord—initiated the Taliban movement. Between 1994 and 1998, in Kandahar and Kabul, I interviewed nearly all the students who were the founding members of the Taliban and the three men she names were never mentioned, except as intermittent financiers. The founders of the Taliban were pious, conservative, simple young villagers who had fought the Soviets as foot soldiers and were now deeply disillusioned with their former leaders for fighting a civil war. They came together to rid Kandahar of criminal gangs. They then travelled around the country asking warlords to help end the civil war and bring peace. When that failed they decided to launch their own movement.

Contrary to Gall's account that they wanted power over Afghanistan from the first, the Taliban founders initially had only three aims—to end the civil war, disarm the population, and introduce an Islamic system. Until they reached the gates of Kabul in late 1995, they had no intentions of ruling the country. Instead they were demanding a Loya Jirga, or meeting of tribal elders, to decide who should rule. Some, like Mullah Borjan, were actually royalists who wanted to call back the former King Zahir Shah from exile. Gall says Borjan was killed at the behest of the ISI in 1996, although it is widely accepted that he died a year earlier in the first attack on Kabul.

All the founding members of the Taliban I interviewed gave a different account from Gall's of the rise of their leader Mullah Mohammed Omar. They all had equal status, the requisite piety, and a strong record of fighting the Soviets. There was no natural commander among them. After much debate they picked Omar as the first among equals, the most pious and apparently the most humble. His status rose only after he insisted that his colleagues swear an oath of allegiance to him. He continues to be powerful. Too much of Gall's information and analysis on the history of the Taliban seems to reflect the views of the Afghan intelligence service, whose own interpretation is flawed and one-sided.

Today, with Pakistan torn apart by unprecedented violence and the situation in Afghanistan still precarious, the Pakistani military has strong reasons to change its past policies of sponsoring wars fought by nonstate organizations. Some changes are happening, but only at a glacial pace. Serious reform needs to start at the lowest level of the military, at the schools and colleges from which the army is drawn, where drastic curriculum changes are needed. The ISI needs to be brought under a code of conduct and accountability, particularly with respect to its dealings with violent organizations. Its personnel should be trained in political realism rather than in ideological prejudices. Unless changes in the army can be made more quickly, there is still the danger that this nuclear power could slip into chaos.

PAKISTAN: MILITARY ROLE IN CIVIL ADMINISTRATION

Pakistan's growing dependence on the armed forces to provide a panacea for the crisis in civil administration appears to be sustaining its image of a "praetorian" or "garrison" state. Despite the military leadership professing no desire to rule the country after a decade of democracy, the government has sought military participation in administration, resulting in the creation of an armed bureaucracy since last year. This trend is evident from increasing military involvement in wide-ranging administrative activities, from managing essential services and monitoring state-owned schools to conducting the census and building non-military roads. Today, the military, under democratic governance, has a wider and deeper participation in civil administration than it had during the martial law regimes. As a result, two processes are simultaneously taking shape: the militarisation of civil society and the civiliaisation of the military community.

Praetorianism implies the exercise of independent political power by a military community which can either use or threaten to use the instrument of violence against the citizenry. In a sense, the military evolves into an entity to manage threats to legitimate authority in a country. In reality, this is possible only in underdeveloped countries which are characterised by poverty, illiteracy, weak middle class, absence of strong political parties, nascent stages of administrative evolution, and dichotomy between tradition and modernity.

This paper attempts to analyse the role of the Pakistan military in civil administration after a decade of democratic governance. It highlights the difference between the military regime which exercised political power under President Zia-ul-Haq and thereafter under an elected government wherein the armed forces possess only administrative power. The analysis draws a distinction between two types of military leadership based on the concept of "arbitrator" and "ruler" type of praetorian military—one which acquires administrative power and the other that holds political power. The paper will also examine the implications of military involvement in civilian administration taking into consideration both the internal and external dimensions.

The extent and parameters of military involvement in the democratic regime can be best explained by a recent statement of the chief of army staff General Parvez Musharaff. While speaking on February 8, 1999, at Sialkot, he outlined broadly that the Pakistan Army, besides defending the national frontiers, has been helping the government to stabilise various institutions and improving the law and order situation within the country. He said that "the armed forces would continue assisting the government in improving the institutional performance besides aiding civilian administration in its efforts to keep law and order intact."

Today the role of the military in civil administration is quite different from the constitutional crises of the past, simply because in the present context the government requires military participation in running civil administration out of dire necessity to maintain law and order, among various other problems, which is quite in contrast to a military coup d'etat during a constitutional crisis. To that extent, the military involvement in civil administration does not necessarily amount to intervention in the political process like in the past decade. Thus, the expanding involvement of the armed forces in the national mainstream brings into focus the role of the military in a democracy.

In the biggest peace-time mobilisation of the armed forces, a quarter million military personnel were employed to conduct the fifth population census in March 1998. Almost 30,000 personnel have been deputed to manage the Water and Power Development Authority (Wapda). The setting up of controversial military courts, which are not as manpower intensive as the involvement with other civilian administrative activities, symbolises the power of the military over the citizens. In April 1998, some 20,000 military personnel were used to investigate "ghost" schools in Punjab which were involved with siphoning education funds into the pockets of politicians and officials. Today, while senior military officers have been appointed governors and chief executives of public sector corporations, their retired counterparts have contested elections and become political leaders. Here a problem could arise on account of the strong levels of bonding or espirit de corps that is traditionally associated with the military community and evident from their personal associations persisting much after their life in uniform. Given this military ethos, there is a danger that a nexus could develop between serving military officers and their retired counterparts, both of whom may be holding positions of power. And this could result in a tacit understanding between the military brotherhood to concentrate power in their hands thereby denying the same to the bureaucrat and politician.

Another disturbing trend is to utilise the military for civilian administrative work like conducting the census, managing the service sector, meting out justice and monitoring state-owned schools. The third trend relates to the incorporation of more military officers into the civil bureaucracy, mainly the federal secretariat, the intelligence, and the police. The military, therefore, runs the country at three levels—at the apex through political appointments like governors, with administrative appointments to manage various public sector corporation, and in the district administration at the second level and at the lowest level by getting the soldiers and sailors to actually perform work professionally not assigned to them. The military, therefore, exercises control in a province from the top, in the district administration at the middle-level through the majors and colonels, and at the grassroots level through its soldiers and sailors who have a ground level presence owing to the various activities assigned to them.

The ubiquitous presence of military officers in civilian administration dilutes democratic culture which upholds values like freedom of the press and transparency in the process of administration, including institutions like the judiciary. On the other hand, the military which has a different ethos essentially oriented to war, tends to downplay these democratic values. To illustrate the point, let us assume that martial law replaces democracy in a country with a powerful press which plays the traditional role of a watchdog and exposes the misdeeds of the political leadership. Thereafter, in the new military regime, the press would not be able to perform an effective part in governance. This is because military leaders have limited experience and exposure to dealing with the press as news reportage on defence affairs is restricted business in developing countries. More specifically, senior military officers do not encourage or entertain searching questions about the state of the armed forces as a rule and tend to give evasive responses when confronted with awkward issues.

Therefore, military leaders functioning as martial law administrators are likely to allow such inherited attitudes to persist even when dealing with citizens outside the military. The military command structure does not encourage liberty and equality among its members who have a well-defined hierarchy only in order to fulfill its professional objectives. To that extent, martial law regimes and democracies are mutually incompatible by nature. This, therefore, results in the militarisation of society which is an undesirable and unhealthy trend for the development of civil societies.

For instance, during 1992-93, senior Pakistan military officers were holding around 100 coveted civilian appointments. This trend of armed forces "capturing" power outside military organisations led to heartburn among civilian officers/technocrats who had aspired for these positions all through their careers. It is relevant to refer to the case of a senior officer in the Pakistan National Shipping Corporation (PNSC) who filed a constitutional petition in the Sindh High Court against Vice Admiral Mansurul Haq of the Pakistan Navy who was the PNSC chairman in May 1993. The issue emerged following a conflict between the Nawaz Sharif regime and the Pakistan Navy over privatisation of the PNSC. While Nawaz Sharif favoured privatisation, the naval lobby oppossed the move. Such a situation clearly underlines how the military lobby's vested interests have a tendency to override national economic interests.

Similarly, excessive involvement of the military in terms of time and manpower in non-military matters is not advisable for the armed forces. The unambiguous role of the armed forces is to wage war against an external enemy. Therefore, involvement in other unrelated spheres of administration only damages the discipline inculcated among the personnel through a particular style of training and functioning. For instance, the extended employment of armed forces for aid to civil power to maintain law and order will bring military personnel into direct contact with citizens and create scope for corruption and misuse of power. It erodes the military culture of discipline. Making the military perform a role it was not trained and oriented for only distorts the armed forces organisationally and leads to civilianisation of the military community.

CIVIL-MILITARY RELATIONS

The new Afro-Asian nation states which became independent republics in the post-World War II, period inherited weak administrative systems with the withdrawal of colonial powers from their countries but had comparitively better organised militaries. These militaries had participated in the World War II and were compelled to develop into professional fighting organisations. The militaries were the only viable option to elected governments, which were ineffective without an appropriate administrative apparatus, as they possessed stronger organisational structures and the most modernised institutions in those societies.

Among the post-colonial societies, Pakistan was created in 1947 as a territorial expression of Muslim nationhood comprising five distinct ethnic streams: Punjabis, Balochis, Bengalis, Pathans and Mohajirs which together did not contribute to a unifying national character. These diverse communities did not help in national integration and instead sharpened the socio-ethnic cleavages which further weakened the social fabric in the country. In turn, these realities coupled with the fact that the Muslim League, which spearheaded the movement for Pakistan, did not have a strong political base in that country, enabled the bureaucracy to fill in the vacuum created by the withdrawal of colonial power and such a situation did not encourage the evolution of political parties or personalities.

The bureaucracy was the de facto ruler in Pakistan from 1951 after political leadership proved to be ineffective. After a decade of bureaucratic rule with the farce of political leadership, the bureaucracy became a corrupt entity. The levels of corruption led to the intervention of the military in politics, resulting in the first military dictatorship of Field Marshal Ayub Khan. Initially, the civil service used the political leadership as an agent to legitimise its rule but with the advent of military rule in 1958, the bureaucracy then switched this role to the armed forces. This resulted in a nexus between the military and the bureaucracy to consolidate power among themselves in order to avoid sharing it with political parties.

The two reasons for the defunct state of civil administrative machinery are corruption and political interference. The administrative bureaucracy primarily comprises the Civil Service of Pakistan (CSP), the Police Service of Pakistan (PSP). It was characterised by its regulatory rather than facilitatory nature. This encouraged corruption in all forms and at all levels in the bureaucracy. For instance, a district magistrate's power could be used to either support or stymie a politician in the initial stages of contesting the elections. The administrative leadership was also able to mislead and influence the public towards a particular course of action given the level of illiteracy in the country. The Civil Service of Pakistan or the upper crust of the bueaucracy commanded great influence over the business community in terms of granting licences and contracts. Eventually during the early 1950s, corruption had reached gargantuan dimensions and 500 cases were registered but the number of convictions were few. The Pakistan Times stated in an editorial:

However depressing it may be, the fact must be faced that, during recent years, a stage had been reached where honest men in the administration or public life rather than the corrupt were regarded as oddities. No branch of administration could claim to be free of the curse; and from the chaprasis and petty clerks to the highest paid officials—and there are many who did not join the scramble for illegal gains—found it most embarrasing to work in this atmosphere, and they were sometimes even discriminated against for refusing to play the politician's dirty game. A section of the Pakistani civil servants has always imagined that Pakistan was created solely for their personal benefit and, not satisfied with accelerated promotion and inflated salaries, they sought to amass wealth by every fair or foul means.

The situation had deteriorated to such an extent that the central government promulgated the Civil Services (Prevention of Corruption) Rules in 1953. The civil servants were corrupt and grouped into various factions; as a result the CSP did not emerge into a strong institution. The peak period of rampant corruption was between 1962 to 1968 with civil service officer's wives also getting involved, and ministers were unable to act against them as they themselves were compromised or powerless.

President Ayub Khan failed to check corruption and this largely resulted in his downfall. After General Yahya Khan took over, he sacked 303 senior civil servants including 38 CSP officers and 16 PSP officers. Thereafter, Z.A. Bhutto, as prime minister, sacked 1,400 civil service officers and introduced political interference in running the bureaucracy. He started the lateral entry scheme for direct recruitment into the bureaucracy with mid-career experience. Critics opine that this scheme was initiated to enable Bhutto to infiltrate the bureaucracy with individuals loyal to him.

President General Zia-ul Haq terminated the lateral entry system into the civil service but replaced this with the institutionalised induction of military officers. Otherwise, he had a cordial relationship with the bureaucracy and all key portfolios like those of secretaries of finance, information, defence, establishment and interior were held by civil servants.

The Pakistan military's credibility stems from the fact that it delivered the goods by way of protecting territorial integrity against India and Afghanistan during the initial phase of nationhood. Field Marshal Ayub Khan, has stated in his biography Friends Not Masters, "Pakistan's survival was vitally linked to the establishment of a well-trained, well-equipped and well-led army." The army was entrusted with wide and varied duties in 1947. Initially, "it had to assist the depleted ranks of the civil administration in maintaining law and order"; secondly, "there were gigantic problems of extrication, protection, movement and administration of millions of refugees from India"; thirdly, "it was entrusted with the task of protecting "the Hindus and Sikhs migrating to India"; lastly, it had to protect the new, ill-defined, lengthy and sensitive borders.

The bureaucracy in general welcomed the prime ministership of Benazir Bhutto and associated her group with younger age profiles to initiate a change in the style of governance. Owing to the anti-military ethos amongst the bureaucracy, as the military and civil service were rivals in the power structure, the civil servants felt an affinity for the Pakistan People's Party (PPP) and the new prime minister. However, during the 20-month PPP regime there was unprecedented glasnost in governance through the medium of inspired leaks about government affairs to newspapers. These leaks could be attributed to a bureaucracy disgruntled with the government-of-the-day.

The conflictual relationship between the PPP and the bureaucrats had three major reasons which merit elucidation. The PPP government had been in Opposition for a decade and, therefore, had an inherent distrust of the bureaucracy. The PPP regime attempted to "short-circuit" the system and inducted party loyalists in an arbitrary manner which disturbed the decision making hierarchy. The PPP politicians were corrupt and their ministers were only involved in making money.

Therefore, the Pakistan military, being the only professional institution remaining in the country, managed to insulate itself from political interference and corruption in relative terms. Hence, the government had no option but to utilise the military, which continues to be the last bastion of excellence, for a host of civilian administrative duties which otherwise in normal circumstances are civil service functions.

While a military takeover of political power in Pakistan amounts to "hard" rule, the armed forces participation in civil administration symbolises "soft" rule. The "soft" rule operates at three levels: the large scale employment of non-commissioned officers, on a relatively smaller scale at the middle management and at the apex with senior officers holding two-star/three ranks as chief executive officers/administrative appointments. Pakistan's experiments with military rule persist as the generals are attempting to gain greater control over civilian institutions during the past year.

Despite a decade of democratic governance, administrative agencies have not been adequately strengthened and the military continues to be the only institution the leadership can use as a last resort. Today, the level of military involvement in various spheres of the national mainstream only indicates that martial law has ceased only on paper, not in reality. The only difference is that military leaders earlier grabbed power from politicians when political, economic and social disorder prevailed but they have now accepted a political directive to support the government in running the day-to-day affairs of the nation. While the military has not taken over the reins of the country for over a decade, after the demise of General Zia-ul Haq in August 1988, it is now being compelled to get seriously involved with civilian administrative duties.

2

Inter-Services Intelligence

INTER-SERVICES INTELLIGENCE ACTIVITIES IN AFGHANISTAN

The Inter-Services Intelligence (ISI) has been heavily involved in covertly running the military intelligence programs in Afghanistan since before the Soviet invasion of Afghanistan in 1979. In the 1980s, the *ISI* systematically coordinated the distribution of arms and financial means provided by the United States' Central Intelligence Agency (CIA) to some factions of the Afghan mujahideen such as the *HeI* of Gulbuddin Hekmatyar. After the Soviet retreat, the ISI and the Pakistan government led by Prime Minister Benazir Bhutto became primary source of supporting the Hekmatyar in his 1992–1994bombardment campaign against the Afghan government and the capital Kabul.

It is widely agreed that after Hekmatyar failed to take over power in Afghanistan, the *ISI* helped to found the Afghan Taliban. The *ISI* in conjunction with other parts of the Pakistan military subsequently provided financial, logistical, military and direct combat support to the Taliban until the attacks of 9/11. It is widely acknowledged that the *ISI* has given the Afghan Taliban safe havens inside Pakistan and supported the Taliban's resurgence in Afghanistan after 9/11 helping them, especially the Haqqani network, carry out attacks inside Afghanistan. Pakistani officials deny this accusation. Allegations have been raised by international government officials, policy analysts and even Pakistani military officials that the ISI in conjunction with the military leadership has also provided some amount of support and refuge to al-Qaeda. Such allegations were increasingly issued when Al-Qaeda leader Osama Bin Laden was killed.

HEZB-E ISLAMI GULBUDDIN

In 1979 the Soviet Union invaded Afghanistan. The ISI and the CIA worked together to recruit Muslims throughout the world to take part in Jihad against the occupying forces. However the CIA had little direct contact with the Mujahideen as the ISI were the main contacts and handlers and they favoured the most radical of the groups, namely the Hezb-e Islami of Gulbuddin Hekmatyar.

In 1991, after the Soviets had left Afghanistan, the ISI tried to install a government under Gulbuddin Hekmatyar with Jalalabad as their provisional capital, but failed. The Afghan Interim Government, which they wanted to install, had Gulbuddin Hekmatyar as Prime Minister and Abdul Rasul Sayyaf as Foreign Minister. The central organizer of the offensive on the Pakistan side was Lieutenant-General Hamid Gul, Director-General of the ISI. The Jalalabad operation was seen as a grave mistake by other mujahideen leaders such as Ahmad Shah Massoud and Abdul Haq. Neither Massoud nor Haq had been informed of the offensive by the ISI beforehand and neither had participated, as both commanders were considered too independent.

After operations by the Shura-e Nazar of Ahmad Shah Massoud and the defection of the communist general Abdul Rashid Dostum and the subsequent fall of the communist Mohammad Najibullah-regime in 1992, the Afghan political parties agreed on a peace and power-sharing agreement (the Peshawar Accords). The Peshawar Accords created the Islamic State of Afghanistan and appointed an interim government for a transitional period to be followed by general elections. According to Human Rights Watch:

The sovereignty of Afghanistan was vested formally in the Islamic State of Afghanistan, an entity created in April 1992, after the fall of the Soviet-backed Najibullah government.... With the exception of Gulbuddin Hekmatyar's Hezb-e Islami, all of the parties... were ostensibly unified under this government in April 1992.... Hekmatyar's Hezb-e Islami, for its part, refused to recognize the government for most of the period discussed in this report and launched attacks against government forces and Kabul generally.... Shells and rockets fell everywhere.

Gulbuddin Hekmatyar received operational, financial and military support from Pakistan. Afghanistan expert Amin Saikal concludes in *Modern Afghanistan: A History of Struggle and Survival*:

Pakistan was keen to gear up for a breakthrough in Central Asia.... Islamabad could not possibly expect the new Islamic government leaders... to subordinate their own nationalist objectives in order to help Pakistan realize its regional ambitions.... Had it not been for the ISI's logistic support and supply of a large number of rockets, Hekmatyar's forces would not have been able to target and destroy half of Kabul.

By 1994, however, Hekmatyar had proved unable to conquer territory from the Islamic State. Australian National University Professor William Maley writes, "in this respect he was a bitter disappointment to his patrons."

Present

Since the Afghan Presidential Elections in late 2009 Afghan President Hamid Karzai has increasingly become isolated surrounding himself with members of Gulbuddin Hekmatyar's Hezb-e Islami. The *Associated Press* reports: "Several of Karzai's close friends and advisers now speak of a president whose doors have closed to all but one narrow faction and who refuses to listen to dissenting opinions."

Al-Jazeera wrote in early 2012 that Presidential Chief of Staff, Karim Khoram from Gulbuddin Hekmatyar's Hezb-e Islami, besides controlling the Government Media and Information Center, enjoys a "tight grip" over President Karzai. Former co-workers of Khoram have accused him of acting "divisive internally" and having isolated Hamid Karzai's "non-Pashtun allies". *Al-Jazeera* observes: "The damage that Khoram has inflicted on President Karzai's image in one year - his enemies could not have done the same." Senior non-Hezb-e Islami Pashtun officials in the Afghan government have accused Khoram of acting as a spy for Pakistan's Inter-Services Intelligence.

AFGHAN TALIBAN

The Taliban were largely founded by Pakistan's Interior Ministry under Nasrullah Barbar and the Inter-Services Intelligence (ISI) in 1994. In 1999, Nasrullah Baber who was the minister of the interior under Bhutto during the Taliban's ascent to power admitted, "we created the Taliban".

William Maley, Professor at the Australian National University and Director of the Asia-Pacific College, writes on the emergence of the Taliban in Afghanistan:

"In 1994, with the failure of [Gulbuddin Hekmatyar's alliance] attempt to oust [the Afghan] Rabbani [administration], Pakistan found itself in an awkward position. Hekmatyar had proved incapable of seizing and controlling defended territory: in this respect he was a bitter disappointment to his patrons.... In October 1994, [Pakistani interior minister] Babar [took] a group of Western ambassadors (including the US Ambassador to Pakistan John C. Monjo) to Kandahar, without even bothering to inform the Kabul government, even though it manned an embassy in Islamabad.... On 29 October 1994, a convoy of trucks, including a notorious ISI officer, Sultan Amir... and two figures who were later to become prominent Taliban leaders, entered Afghanistan."

The ISI used the Taliban to establish a regime in Afghanistan which would be favourable to Pakistan, as they were trying to gain strategic depth. Since the creation of the Taliban, the ISI and the Pakistani military have given financial, logistical, military including direct combat support.

According to Pakistani Afghanistan expert Ahmed Rashid, "between 1994 and 1999, an estimated 80,000 to 100,000 Pakistanis trained and fought in Afghanistan" on the side of the Taliban. Peter Tomsen also stated that up until 9/11 Pakistani military and ISI officers along with thousands of regular Pakistani armed forces personnel had been involved in the fighting in Afghanistan.

Human Rights Watch wrote in 2000: "Of all the foreign powers involved in efforts to sustain and manipulate the ongoing fighting [in Afghanistan], Pakistan is distinguished both by the sweep of its objectives and the scale of its efforts, which include soliciting funding for the Taliban, bankrolling Taliban operations, providing diplomatic support as the Taliban's virtual emissaries abroad, arranging training for Taliban fighters, recruiting skilled and unskilled manpower to serve in Taliban armies, planning and directing offensives, providing and facilitating shipments of ammunition and fuel, and... directly providing combat support."

In 1998, Iran accused Pakistani commandos of "war crimes at Bamiyan". The same year Russia said, Pakistan was responsible for the "military expansion" of the Taliban in northern Afghanistan by sending large numbers of Pakistani troops including ISI personnel some of whom had subsequently been taken as prisoners by the anti-Taliban United Front (Northern Alliance).

In 2000, the UN Security Council imposed an arms embargo against military support to the Taliban, with UN officials explicitly singling out Pakistan. The UN secretary-general criticized Pakistan for its military support and the Security Council stated it was "deeply distress[ed] over reports of involvement in the fighting, on the Taliban side, of thousands of non-Afghan nationals." In July 2001, several countries including the United States, accused Pakistan of being "in violation of U.N. sanctions because of its military aid to the Taliban." The Taliban also obtained financial resources from Pakistan. In 1997 alone, after the capture of Kabul by the Taliban, Pakistan gave $30 million in aid and a further $10 million for government wages.

"The Taliban are not Islam - the Taliban areIslamabad. " After the 9/11 attacks, Pakistan claimed to have ended its support to the Taliban. But with the fall of Kabul to anti-Taliban forces in November 2001, ISI forces worked with and helped Taliban militias who were in full retreat. In November 2001, Taliban and Al-Qaeda combatants as well as Pakistani ISI and other military operatives were safely evacuated from the Afghan city of Kunduz on Pakistan Armycargo aircraft to Pakistan Air Force bases in Chitral and Gilgit in Pakistan's Northern Areas in what has been dubbed the "Airlift of Evil"

A range of officials inside and outside Pakistan have stepped up suggestions of links between the ISI and terrorist groups in recent years. In fall 2006, a leaked report by a British Defense

Ministry think tank charged, "Indirectly Pakistan (through the ISI) has been supporting terrorism and extremism—whether in London on 7/7 [the July 2005 attacks on London's transit system], or in Afghanistan, or Iraq." In June 2008, Afghan officials accused Pakistan's intelligence service of plotting a failed assassination attempt on President Hamid Karzai; shortly thereafter, they implied the ISI's involvement in a July 2008 Taliban attack on the Indian embassy. Indian officials also blamed the ISI for the bombing of the Indian embassy. Numerous U.S. officials have also accused the ISI of supporting terrorist groups including the Afghan Taliban. U.S. Defense Secretary Robert Gates said "to a certain extent, they play both sides." Gates and others suggest the ISI maintains links with groups like the Afghan Taliban as a "strategic hedge" to help Islamabad gain influence in Kabul once U.S. troops exit the region. U.S. Chairman of the Joint Chiefs of Staff Admiral Mike Mullen in 2011 called the Haqqani network (the Afghan Taliban's most destructive element) a "veritable arm of Pakistan's ISI". He further stated, "Extremist organizations serving as proxies of the government of Pakistan are attacking Afghan troops and civilians as well as US soldiers."

From 2010, a report by a leading British institution also claimed that Pakistan's intelligence service still today has a strong link with the Taliban in Afghanistan. Published by the London School of Economics, the report said that Pakistan's ISI has an "official policy" of support for the Taliban. It said the ISI provides funding and training for the Taliban, and that the agency has representatives on the so-called Quetta Shura, the Taliban's leadership council. The report, based on interviews with Taliban commanders in Afghanistan, was written by Matt Waldman, a fellow at Harvard University. "Pakistan appears to be playing a double-game of astonishing magnitude," the report said. The report also linked high-level members of the Pakistani government with the Taliban. It said Asif Ali Zardari, the Pakistani president, met with senior Taliban prisoners in 2010 and promised to release them. Zardari reportedly told the detainees they were only arrested because of American pressure. "The Pakistan government's apparent duplicity – and awareness of it among the American public and political establishment – could have enormous geopolitical implications,"

Waldman said. "Without a change in Pakistani behaviour it will be difficult if not impossible for international forces and the Afghan government to make progress against the insurgency." Amrullah Saleh, director of Afghanistan's intelligence service until June 2010, told Reuters in 2010 that the ISI was "part of a landscape of destruction in this country".

In March 2012, the Commander of NATO forces in Afghanistan, General John Allen, told the United States Senate that as of 2012 there was still no change in Pakistan's policy of support for the Afghan Taliban and its Haqqani network. When asked by U.S. Senator John McCain whether the ISI had severed its links with the Afghan Taliban, General Allen testified: "No."

Recruitment

The Pakistani army through ISI have been accused of recruiting fighters and suicide bombers for the Afghan Taliban among the 1.7 million registered and 1-2 million unregistered Pashtun Afghan refugees living in refugee camps and settlements along the Afghan-Pakistan border in Pakistan many of whom have lived there since the Soviet war in Afghanistan.

Abdel Qadir, an Afghan refugee who returned to Afghanistan, says Pakistan's Inter-Services Intelligence had asked him to either receive training to join the Afghan Taliban or for him and his family to leave the country. He explains: "It is a step by step process. First they come, they talk to you. They ask you for the information.... Then gradually they ask you for people they can train and send [to Afghanistan].... They say, 'Either you do what we say, or you leave the country.'"

Janat Gul, another former refugee who returned to Afghanistan, told the UN Office for the Coordination of Humanitarian Affairs, that Afghan refugees which had been successfully recruited by the ISI were taken to Pakistani training camps which had previously been used during the times of the Soviet war in Afghanistan.

According to an investigative report among Afghan refugees inside Pakistan by the *New York Times*, people testified that "dozens of families had lost sons in Afghanistan as suicide bombers and fighters" and "families whose sons had died as suicide bombers in Afghanistan said they were afraid to talk about the deaths because of pressure from Pakistani intelligence agents, the ISI."

Pakistani and Afghan tribal elders also testified that the ISI was arresting or even killing Taliban members which wanted to quit fighting and refused to re-enlist to fight in Afghanistan or to die as suicide bombers. One former Taliban commander told the *New York Times* that such arrests were then sold to the Westerners and others as part of a supposed Pakistani collaboration effort in the War against Terror.

Provision of Safe Haven

Gen. James L. Jones, then NATO's supreme commander, in September 2007 testified in front of the U.S. Senate Foreign Relations Committee that the Afghan Taliban movement uses the Pakistani city of Quetta as their main headquarters. Pakistan's minister for information and broadcasting, Tariq Azim Khan, mocked the statement by saying, if there were any Taliban in Quetta, "you can count them on your fingers."

Training

From 2002 until 2004, without major Taliban activities, Afghanistan witnessed relative calm with Afghan civilians and foreigners being able to freely and peacefully walk the streets of major cities and reconstruction being initiated. The 2010 testimonies of former Taliban commanders show that Pakistan through its Inter-Services Intelligence was however "actively encouraging a Taliban revival" from 2004-2006.

The effort to reintroduce the Afghan Taliban militarily in Afghanistan was preceded by a two-year, large-scale training campaign of Taliban fighters and leaders conducted by the ISI in several training camps in Quetta and other places in Pakistan. From 2004-2006 the Afghan Taliban consequently started a deadly insurgency campaign in Afghanistan killing thousands of civilians and combatants and thereby renewing and escalating the War in Afghanistan (2001-present). One Taliban commander involved in the Taliban resurgence said that 80 percent of his fighters had been trained in an ISI camp.

HAQQANI NETWORK

The ISI have close links to the Haqqani network and contribute heavily to their funding. U.S. Chairman of the Joint Chiefs of Staff Admiral Mike Mullen in 2011 called the Haqqani network (the Afghan Taliban's most destructive element) a "veritable arm of Pakistan's ISI". He further stated:

"Extremist organizations serving as proxies of the government of Pakistan are attacking Afghan troops and civilians as well as US soldiers." —Former U.S. Chairman of the Joint Chiefs of Staff Mike Mullen, *2011*

Mullen said, the U.S. had evidence that the ISI directly planned and spar-headed the Haqqani 2011 assault on the U.S. embassy, the June 28 Haqqani attack against the Inter-Continental Hotel in Kabul and other operations. It is widely believed the suicide attack on the Indian embassy in Kabul was also planned with the help of the ISI A report in 2008 from the Director of National Intelligence stated that the ISI provides intelligence and funding to help with attacks against the International Security Assistance Force, the Afghan government and Indian targets.

Al Qaeda

Besides supporting the Hezb-e Islami of Gulbuddin Hekmatyar, the ISI in conjunction with Saudi Arabia strongly supported the faction of Jalaluddin Haqqani (Haqqani network) and allied Arab groups such as the one surrounding financier Bin Laden, nowadays known as Al-Qaeda, during the war against the Soviets and the Afghan communist government in Afghanistan in the 1980s.

In 2000, British Intelligence reported that the ISI was taking an active role in several Al Qaeda training camps from the 1990s onwards. The ISI helped with the construction of training camps for both the Taliban and Al Qaeda. From 1996 to 2001 the Al Qaeda of Osama Bin Laden and Ayman al-Zawahiri became a state within the Pakistan-supported Taliban state. Bin Laden sent Arab and Central Asian Al-Qaeda militants to join the Taliban's and Pakistan's fight against the United Front (Northern Alliance) among them his Brigade 055.

It is believed that there is still contact between Al-Qaeda and the ISI today. The former Afghan intelligence chief Amrullah Saleh has repeatedly stated that Afghan intelligence believed and had shared information about Osama Bin Laden hiding in an area close to Abbottabad, Pakistan, four years before he (Bin Laden) was killed there. Saleh had shared the information with Pakistani President Pervez Musharraf who had angrily brushed off the claim taking no action.

In 2007, the Afghans specifically identified two Al-Qaeda safe houses in Manshera, a town just miles from Abbottabad, leading them to believe that Bin Laden was possibly hiding there. But Amrullah Saleh says that Pakistani President Pervez Musharraf angrily smashed his fist on a table when Saleh presented the information to him during a meeting in which Afghan President Hamid Karzai also took part. According to Saleh, "He said, 'Am I the president of the Republic of Banana?' Then he turned to President Karzai and said, 'Why have you brought this Panjshiri guy to teach me intelligence?'"

A December 2011 analysis report by the Jamestown Foundation comes to the conclusion that "in spite of denials by the Pakistani military, evidence is emerging that elements within the Pakistani military harboured Osama bin Laden with the knowledge of former army chief General Pervez Musharraf and possibly current Chief of Army Staff (COAS) General Ashfaq Pervez Kayani. Former Pakistani Army Chief General Ziauddin Butt (a.k.a. General Ziauddin Khawaja) revealed at a conference on Pakistani–U.S. relations in October 2011 that according to his knowledge the then former Director-General of Intelligence Bureau of Pakistan (2004–2008), Brigadier Ijaz Shah (retd.), had kept Osama bin Laden in an Intelligence Bureau safe house in Abbottabad." Pakistani General Ziauddin Butt said Bin Laden had been hidden in Abbottabad "with the full knowledge" of Pervez Musharraf. But later Butt denied making any such statement.

Assassination of Pivotal Afghan Leaders

The ISI has been involved in the assassination of major Afghan leaders which have been described as pivotal for the future of

Afghanistan. Among those leaders are the main anti-Taliban resistance leader and National Hero of Afghanistan Ahmad Shah Massoud and the prominent Pashtun anti-Soviet and anti-Taliban resistance leader Abdul Haq. The ISI has also been accused of having been involved in the murder of former Afghan president and chief of the Karzai's administration High Peace Council Burhanuddin Rabbaniand several other anti-Taliban leaders.

In the case of Ahmad Shah Massoud, who was killed by two Arab suicide bombers two days before September 11 attacks in the United States in 2001 ("9/11"), the Arab suicide assassins - supposed journalists - were granted with multiple entry visas valid for a year in early 2001 by Pakistan's embassy in London. As author and Afghanistan expert Sandy Gall writes such multiple visas for a year are "unheard of for journalists normally". The ISI subsequently facilitated the two men's passage through Pakistan over the Afghan border into Taliban territory. Afghan journalist Fahim Dashty says, "Al-Qaida, the Taliban, other terrorists, the Pakistan security services — they were all working together... to kill him."

Abdul Haq, who was killed by the Taliban on October 26, 2001, enjoying strong popular support among Afghanistan's Pashtuns wanted to create and support a popular uprising against the Taliban - also dominantly Pashtuns - among the Pashtuns. Observers believe that the Taliban were only able to capture him with the collaboration of the ISI.

Ahmad Shah Massoud had been the only resistance leader able to defend vast parts of his territory against the Taliban, Al-Qaeda and the Pakistani military and was sheltering hundreds of thousands of refugees which had fled the Taliban on the territory under his control. He had been seen as the leader most likely to lead post-Taliban Afghanistan. After his assassination, Abdul Haq was seen as one of the main contenders for that position. He had private American backers which had facilitated his re-entry into Afghanistan after 9/11. But journalists reported about tensions between the CIA and Haq. Former CIA director George Tenet reports that, at the recommendation of one of Haq's private American lobbyists Bud McFarlane, CIA officials met with Abdul Haq in Pakistan but after assessing him urged him not to enter

Afghanistan. Both leaders, Massoud and Haq, were recognized for being fiercely independent from foreign especially Pakistani influence. Both, two of the most successful anti-Soviet resistance leaders, were rejecting the Pakistani claim of hegemony over the Afghan mujahideen.

Abdul Haq was quoted as saying during the anti-Soviet period: "How is that we Afghans, who never lost a war, must take military instructions from the Pakistanis, who never won one?" Massoud during the Soviet period said to the Pakistani Foreign Minister who had asked him to send a message to the Russians through the Pakistanis who were conducting talks "on behalf of our Afghan brethren": "Why should I send a message? Why are you talking on our behalf? Don't we have leaders here to talk on our behalf?" Khan replied "This is how it has been and how it will be. Do you have a message?" Massoud told the foreign minister that "nobody who talks on our behalf will have any kind of result." Consequently, neither Massoud nor Abdul Haq were consulted before and neither participated in the Battle of Jalalabad (1989) in which the ISI tried but failed to install Gulbuddin Hekmatyar as the post-communist leader of Afghanistan.

Allegations of Drug Running

It is alleged that the ISI have control over opium production and refining and also control all smuggling operations between Afghanistan and Pakistan. Wendy Chamberlin in testimony before the United States House of Representatives International Relations Committee stated that the ISI involvement in the drug trade was substantial.

INTER-SERVICES INTELLIGENCE ACTIVITIES IN INDIA

The Inter-Services Intelligence (abbreviated as ISI), has been involved in running the military intelligence program in India, with one of the subsections of its Joint Intelligence Bureau (JIB) department devoted to perform various operations in India and related to them. The Joint Signal Intelligence Bureau department has been also supporting Kashmiri militants in regards to communication. The Joint Intelligence North section of the Joint Counter-Intelligence Bureau wing deals particularly with India.

In the 1950s the ISI's Covert Action Division supplied arms to insurgents in Northeast India.

India has also accused the ISI of reinvigorating terrorism in the country via support to the pro-Khalistan militant groups such as International Sikh Youth Federation, in order to take revenge against India for its help in liberation of Bangladesh as well as to destabilize the Indian State. A report by India's Intelligence Bureau indicated that ISI was "desperately trying to revive Sikh" militant activity in India. The ISI is also allegedly active in printing and supplying counterfeit Indian rupeenotes.

History

The ISI was created after the Indo-Pakistani War of 1947, due to Military Intelligence of Pakistan's weak performance. When Zia-ul-Haq seized power in July 1977, he started his K2 (Kashmir and Khalistan) strategy, initiating Operation Tupac. He gave ISI the duty to make Jammu and Kashmir a part of Pakistan, and to send terrorists to Punjab. According to arrested ISI agents, the intelligence agency's aims are to confound Indian Muslims using Kashmiri Muslims, extend the ISI network in India, cultivate terrorists and terrorist groups, cause attacks similar to the 1993 Bombay bombings in other cities, and create a state of insurgency in Muslim-dominated regions. The ISI has allegedly set up bases in Nepal and Bangladesh, which are used for operations in North-East India.

OPERATIONS IN JAMMU AND KASHMIR

About Rs. 24 million are paid out per month by the ISI, in order to fund its activities in Jammu and Kashmir. Pro-Pakistani groups were reportedly favoured over other militant groups. Creation of six militant groups in Kashmir, which included Lashkar-e-Taiba (LeT), was aided by the ISI. According to American Intelligence officials, ISI is still providing protection and help to LeT. The Pakistan Army and ISI also LeT volunteers to surreptitiously penetrate from Pakistan Administrated Kashmir to Jammu and Kashmir. As of 2010, the degree of control that ISI retains over LeT's operations is not known. The LeT was also reported to have been directed by the ISI to widen its network in the Jammu region where a considerable section of the populace comprised Punjabis.

INVOLVEMENT IN TERRORIST ATTACKS

Involvement with 26/11 Attacks

Zabiuddin Ansari, a Lashkar-e-Taiba militant accused for his involvement in the 2008 Mumbai attacks, said that ISI and Pakistani army officials were involved in planning the attacks and had attended the meetings. An Indian report, summarising intelligence gained from India's interrogation of David Headley, alleged that ISI had provided support for the attacks by providing funding for reconnaissance missions in Mumbai. The report included Headley's claim that Lashkar-e-Taiba's chief military commander, Zaki-ur-Rahman Lakhvi, had close ties to the ISI. He alleged that "every big action of LeT is done in close coordination with [the] ISI."

Involvement with Mumbai Train Blasts

ISI was alleged of planning the 2006 Mumbai train bombings and the Indian government said that the ISI, LeT and SIMI planned the attacks, but a former Indian home ministry officer submitted his declaration in the Supreme Court of India which said that he was told by a former member of the CBI-SIT team that both the terror attacks (Parliament and Mumbai) were staged "with the objective of strengthening the counter-terror legislation(sic).

Counterfeit Indian Rupee Notes

The ISI has been alleged to print counterfeit Indian rupee notes, which are believed to be printed in Muzaffarabad. In January 2000, the Nepal police raided Wasim Saboor's house, who was an official of the Pakistani embassy of Kathmandu. They found fifty thousand Indian rupee notes, each of Rs. 50 denomination.

REASONS WHY THE US AND INDIA DEMONIZE PAKISTAN'S ISI

Pakistan's Inter Services Intelligence agency, or ISI as it is popularly known, is seen as their nemesis by those who have tried to undermine the security interests of the country one way or the other. It is no wonder then that in past few years the Americans unleashed a strong ISI-bashing campaign, with India following suit.

The Americans made no bones about their dislike for this agency, blaming it for working against their interests in Afghanistan. The Indians also see an ISI agent behind every rock in Kashmir and in Afghanistan where they are trying to dig their heels. They do not hesitate to pin on ISI the blame for the freedom struggle in Kashmir or for acts of terrorism by Indian extremists. Until recently the Karzai government dominated by the anti-Pakistan Northern Alliance also remained hostile to ISI.

Not too long ago, under intense American pressure the weak Zardari government made an unsuccessful attempt at neutralizing and subduing this agency in disregard to the existing sensitive regional security environment, by moving it out of the army control and placing it under the controversial and embattled Zardari loyalist interior minister – Rehman Malik. This did not succeed for a simple reason. The role of ISI as the eyes and ears of the Pakistan's military – the bedrock of country's security, is critical particularly at a time when the country faces multiple threats to its security.

Washington's darling in the Afghan-Soviet War

Ironically, this is the same ISI that was Washington's darling during the 1980s when it was master minding the jihad against invading Soviet forces in Afghanistan. The role that ISI then played was congruent with American interests. The defeat of the Soviet Union would have meant realization of an American dream – avenging the humiliation of Vietnam. They held ISI in high esteem for its competence and professionalism and gladly funnelled arms and funds to the Afghan mujahedeen through it. The ISI strategized the resistance and organized and trained the mujahedeen fighters, working in close collaboration with the CIA and the mujahedeen leaders, forcing the Soviets to retreat.

But as soon as the Americans had negotiated a quid pro quo – Russian withdrawal from South America in exchange for safe

Soviet exit from Afghanistan, they disappeared in the middle of the night leaving Afghanistan in a quandary. The political turmoil that followed created chaos and instability owing to the failure of mujahedeen leadership, presenting as a result a security nightmare for Pakistan.

Taliban-US-Pakistan Relations and the Indian Threat

In this chaos a group of young Afghan religious students, many of them former fighters from the resistance, calling themselves Taliban (in Pushto language Taliban means students), swept through the country with popular support to establish their rule. Interested to keep their presence alive, the Americans maintained contacts and supported them, ignoring their orthodox beliefs, their harsh rule and even the presence of Al Qaeda in their midst. This continued until it was time for the Americans to overthrow their government in order to serve the changing American interests.

While the Taliban government was in control, Pakistan too maintained friendly relations with them in the interest of keeping its western border secure, extending whatever support it could. The ISI played a role through the contacts it had developed during war against the Soviets.

In the wake of 9/11 things began to change. Having invaded Afghanistan in the name of war on terror, branding Taliban as brutes and their resistance as terrorism, the Americans wanted the Pakistan army and the ISI to join the war.

This posed a serious security concern for Pakistan. It could destabilize the Pak-Afghan border and strain relations with the Pashtun tribes on both sides of the Durand Line, the British drawn boundary that cut through the Pashtun region to divide British India and Afghanistan and which Pakistan had inherited. The fact that Pakistan's border region, called Federally Administered Tribal Areas (FATA) is autonomous where the writ of the Pakistan Government does not prevail made matters more complex.

Pakistan's military doctrine is based primarily on meeting the main threat from India on its eastern border while maintaining a peaceful border with Afghanistan in the west. A direct conflict with the Taliban would have forced Pakistan to divert its military assets from eastern to the western front, thus thinning out its defenses against India. This was the last thing Pakistan wanted to do because of its unfavourable ratio of 1:4 against India in terms of conventional forces. Understandably, President Musharraf was unwilling to do the American bidding.

U.S. PROJECTION OF ITS MILITARY FAILURES ONTO PAKISTAN

There always is a problem with powers that begin to act in imperialistic fashion. Their vision of the world becomes coloured. They tend to believe that pursuit of their imperialist designs takes precedence over the national interests of those who cannot stand up to them, even if that means compromising their own national and security interests. America had also been behaving as one such imperial power and treated its smaller allies more like colonies. President Musharraf was threatened that in case of noncompliance with America's wishes, "Pakistan would be bombed into the stone-age". Musharraf was coerced into conceding to American demands.

Despite the state-of-the-art surveillance equipment and military hardware, the US and NATO forces failed to stop the Taliban fighters from moving back and forth into the unmarked Pak-Afghan border that passes through a treacherous mountainous region to regroup and strike on the invading foreign troops.

The American commanders reacted by demanding that the Pakistan army engage these fighters and seal the border. Those with even the slightest knowledge of the area would know that the Americans were asking for the moon. This was physically impossible.

Pakistan army's operations failed. In the process it earned a severe backlash from the local tribes who resented army's action against their kinsmen from across the border who sought refuge in their area, as it violated the old tribal custom of providing sanctuary to any one who asked for it, even it was an enemy. The Pakistan army paid a heavy price. More soldiers died in this action than the combined number of casualties that the US and NATO troops have suffered in Afghanistan so far.

President Musharraf under advice of his army commanders and the intelligence community called off the action and resorted to persuasion instead. Through jirgas (assembly of tribal elders) effort was made for the tribesmen to voluntarily stop the influx of Taliban fighters. It didn't succeed either. This was not to the liking of the American commanders. They blamed the ISI for working against their interests.

Washington Accuses the ISI of Complicity with Insurgents

Washington and the American media frequently alleged that elements within ISI were maintaining contacts with the Taliban and attributed the failure of American troops in combating the Taliban to these contacts. Such allegations were also found to be part of the raw, unverified and even fabricated field reports 'leaked' in Afghanistan recently and splashed in the western media. The Americans have in the past also described the ISI to be out of control and demanded of the Pakistan government to purge the agency of Taliban sympathizers.

This is ridiculous. Firstly, ISI is a military organization operating under strict organizational control and discipline where officers are rotated in the normal course. It functions according to a defined mandate, unlike armed forces in some other countries and unlike the CIA which is known to be an invisible government on its own. Above all, Pakistan and its military are committed to weeding out religious extremism as a matter of state policy.

Secondly, if the American troops are so incapable of overcoming a rag tag army of Taliban and if the complicity of ISI with the Taliban can be instrumental in changing the course of the American war, then it is a sad day for America as a super power and the strength of NATO forces becomes questionable.

Thirdly, in the world of intelligence, contacts are kept even with the enemy and at all times. CIA keeps contacts within Russia and other hostile countries. Israel, the great American ally, spies on America itself. It is common for all intelligence agencies to do this in the security interests of their countries. Why then should America expect an exception to be made in case of ISI? Why should contacts that ISI developed with the mujahedeen and the

Taliban earlier, and which if it does still maintain, become a source of such great concern for the American administration?

It is strange that America expects ISI to serve the American agenda instead of Pakistan's interests first. One cannot forget that the Americans have a long history of abandonment of friends and allies and when they repeat this in Afghanistan citing their own national interest, despite their promises to the contrary, why should Pakistan be expected to be caught with pants down? Why Pakistan's military and the intelligence agency should be expected to abdicate their duty and not do what is necessary to ensure Pakistan's security in the long term?

It has often been argued that America expects Pakistan to be actively engaged in the Afghan war in return for the military assistance it provides. The answer is quite simple. The American establishment is doing all that needs to be done in support of its own war and not for the love of Pakistan. The war is theirs, not Pakistan's. Pakistan should do and is doing what is necessary and feasible, without jeopardizing its own security.

As for the assistance, bulk of the $10 billion that America gave in the past and was branded as "aid" was in fact the reimbursement of expenses that Pakistan had already incurred in supporting the war effort. The rest was to meet Pakistan's needs for operations in the border areas and for fighting terrorism that arose out of the war. The Americans still owe $35 billion to reimburse the losses Pakistan has incurred due to this war. As for the F16s that Pakistan is getting from the US, it pays for them, despite strict restrictions over their usage.

THE INDIAN-ISRAELI ATTEMPT TO DESTABILIZE PAKISTAN

While Americans had their issues with ISI, the Indians and Israelis began having their own. The agency exposed the growing Indian and Israeli confluence in Afghanistan to destabilize Pakistan. This happened right under the nose of the Americans and obviously not without their knowledge and consent. India having deployed its troops in the name of infra-structure development in league with Karzai government and with American funding and having established seven consulates along the sparsely populated Pak-Afghan border was engaged in heavily bribing the influential but ignorant and susceptible tribal leaders to spread disaffection among the local tribesmen against Pakistan.

Evidence was also unearthed by ISI about how the Indians bought the loyalties of Tehrik-e-Taliban Pakistan (TTP), a grouping of Pakistani tribesmen from FATA and Uzbek fighters from previous wars who settled in the region. The TTP were influenced by the same orthodox religious beliefs as the Taliban in Afghanistan and were active in propagating them in their own areas. They were recruited to launch terror activities in the urban centers of Pakistan, including the capital Islamabad, and were funded, trained and equipped in Afghanistan jointly by the Indian, Israeli and Afghan intelligence agencies. A group from amongst them managed to gain control of Swat area adjoining FATA through coercion of the local population, which was later cleared by the Pakistan army after a major surgical intervention.

The ISI also laid bare strong physical evidence of Indian involvement in supporting insurgency in Balochistan by way of funding, training and equipping misguided and disgruntled Baloch elements grouped under various names including the Balochistan Liberation Army that was led by the fugitive grandson of the notable Bugti tribal chief – Akbar Bugti. His comings and goings in the Indian consulate at Kandahar and the Indian intelligence HQ in Delhi were photographed and his communications intercepted. Numerous training camps in the wilderness of Balochistan were detected where Indian trainers imparted training in guerilla warfare and the use of sophisticated weapons, which otherwise could not be available to the Baloch tribesmen. Flow of huge funds from Afghan border areas to the insurgents was detected that was traced back to the Indian consulates.

Summary and Conclusion

The objective of the TTP, and behind the scene that of the Indians and the Israelis, was to make the world believe that Pakistan was under threat of capitulating to terrorist and insurgent elements who were about to take control of Pakistan's nuclear assets. Their goal: to denuclearize Pakistan through foreign intervention. These efforts have not succeeded. Undoubtedly, the army and the ISI played a crucial role in foiling the plots of subversion in Balochistan and the Pashtun region and exposing the foreign hands involved, including those of CIA, RAW, Mossad, RAMA and MI6. Terrorism may not yet be eliminated but Pakistan faces no existential threat.

It should be no surprise to the Americans, Indians and the Israelis if they find in ISI an adversary to reckon with. It is also not surprising that the ISI is in their perception, a rogue organization, for it has stood between them and Pakistan's national security interests. Their frustration and ire, therefore, is understandable.

3

The ISI and Terrorism: In the Wake of the Allegation

INTRODUCTION

Pakistan's military intelligence agency, the Inter-Services Intelligence (ISI), has long faced accusations of meddling in the affairs of its neighbors. A range of officials inside and outside Pakistan have stepped up suggestions of links between the ISI and terrorist groups in recent years. In fall 2006, a leaked report by a British Defense Ministry think tank charged, "Indirectly Pakistan (through the ISI) has been supporting terrorism and extremism—whether in London on 7/7 [the July 2005 attacks on London's transit system], or in Afghanistan, or Iraq." In June 2008, Afghan officials accused Pakistan's intelligence service of plotting a failed assassination attempt on President Hamid Karzai; shortly thereafter, they implied the ISI's involvement in a July 2008 attack on the Indian embassy. Indian officials also blamed the ISI for the bombing of the Indian embassy. Pakistani officials have denied such a connection.

Numerous U.S. officials have also accused the ISI of supporting terrorist groups, even as the Pakistani government seeks increased aid from Washington with assurances of fighting militants. In a May 2009 interview with CBS' *60 Minutes*, U.S. Defense Secretary Robert Gates said "to a certain extent, they play both sides." Gates and others suggest the ISI maintains links with groups like the Afghan Taliban as a "strategic hedge" to help Islamabad gain influence in Kabul once U.S. troops exit the region. These allegations surfaced yet again in July 2010 when WikiLeaks.org made public (*NYT*) a trove of U.S. intelligence records on the war in Afghanistan. The documents described ISI's links to militant groups fighting U.S. and international forces in Afghanistan. In April 2011 during a visit to Pakistan, U.S. Chairman of the Joint Chiefs of Staff Admiral Mike Mullen pointed to ISI's links with one such group, the Haqqani network. The May 1, 2011, killing of America's most wanted terrorist Osama bin Laden in a Pakistani military town not far from Islamabad raised new questions over army and ISI support for the al-Qaeda leader and the legitimacy of their counterterrorism efforts. Pakistan's government has repeatedly denied allegations of supporting terrorism, citing as evidence its cooperation in the U.S.-led battle against extremists in which it has taken significant losses both politically and on the battlefield.

PAKISTAN'S UNHOLY ALLIANCE THE MILITANTS

The heavily guarded, high-walled concrete structure right in the heart of Islamabad has no signboard. But everyone in the capital knows that the sprawling compound surrounded by barbed wire houses the Directorate for Inter-Services Intelligence (ISI). The spy agency has been the country's big brother. It is powerful, ubiquitous and has functioned with so much autonomy from the central government that it has almost become a state within a state. It is not only responsible for intelligence gathering, but also acts as a determinant of Pakistan's foreign policy and a vehicle for its implementation. For every civilian and military government, control of the ISI was seen as crucial to maintaining a firm grip on power. The agency had been so powerful for so long that it played by its own rules. Its various heads had contrasting profiles, but emerged among the most powerful figures in the country's establishment. For years they ran semi-independent operations in Afghanistan and Kashmir and helped to form and topple civilian governments.

For more than two decades the ISI had sponsored Islamic militancy to carry out its secret wars. It was a crucial partner in the CIA's biggest covert operation ever, one that forced the Soviet

Union to pull out of Afghanistan and served as a catalyst to the disintegration of the communist superpower. In Afghanistan, as well as Kashmir, the agency discovered the effectiveness of covert warfare as a method of bleeding a stronger adversary, while maintaining the element of plausible deniability. The ISI falls directly within Pakistan's military chain of command and had also served as an instrument for promoting the military's domestic political agenda and the guardian of its self-professed 'ideological frontiers' of the country. Almost all ISI officers are regular military personnel, who are rotated in and out for a fixed tenure. However, there have been some exceptions. The export of jihad sponsored by the ISI had its blowback. It had allowed the Islamists a huge space for their activities. State patronage, in the form of an 'unholy alliance' between the military and the mullahs, resulted in an unprecedented rise of radical Islam. The ISI had helped to create much of the Islamic militancy and religious extremism that Musharraf was confronted with. That unholy alliance had been a major factor in the country's drift to Islamic fundamentalism. Founded in 1948 by a British army officer, Major-General R. Cawthome, then Deputy Chief of Army Staff in Pakistan, the agency was initially charged with performing all intelligence tasks at home and abroad. Its scope of operation extended to all areas related to national security. Until the 1960s, the ISI largely remained an obscure organization that confined itself to playing its specified role. But in the mid 1970s its scope was expanded to domestic politics. Ironically, it was a civilian leader, Zulfikar Ali Bhutto who created the ISI's internal wing which played a critical role in the ousting of his government a few years later. It was to cast its heavy shadow over the country's politics in later years.

A charismatic and populist leader, Bhutto took over a humiliated military and a brutally truncated Pakistan in search of a new identity, following the Indian-supported secession of East Pakistan (now Bangladesh), in 1971. The country had been dismembered by civil war and Indian military action. It was no longer the country created by Mohammed Ali Jinnah. The Pakistan that emerged from the ashes of defeat required a different geopolitical orientation. A revolt by young Turks in the army in the aftermath of a humiliating defeat in the war had forced the military ruler, General Yahya Khan, 3 to hand over power to Bhutto. His legitimacy was rooted in the country's first democratically held elections in 1970. Bhutto's socialist Pakistan People's Party had swept the fateful polls in the western wing of then-united Pakistan, trouncing the right-wing Islamic parties.

The scion of a feudal family of Sindh and a former foreign minister, Bhutto was seen as the savior of the new Pakistan. He successfully negotiated with India the release of 90,000 troops taken prisoner by Indian forces. He had come to power with the largest support base of any Pakistani leader since the inception of the country in 1947.

But he failed to sustain the confidence of the nation. Bhutto was ideally placed to put into practice the objective of social democracy and develop secular ideas and institutions. But it did not happen. Bhutto's strategy was to bring the disparate elements of this divided society together through a kind of Islamic nationalism, which was then supposed to create the cohesion and stability necessary for socialist economic reforms, but unfortunately all Bhutto succeeded in doing was to rehabilitate religious extremism. Under pressure from the religious parties whose cooperation they were courting, Bhutto's government increased the religious content in school syllabuses and, succumbing to pressure from Saudi Arabia as well as to the demands of religious parties, declared the Ahmedis, an Islamic sect, to be non-Muslim. This apparently minor action had long-term implications for the country as it fuelled Islamic zealotry and sharpened the sectarian divide. Bhutto's attempt to co-opt religious elements merely emboldened them, and eventually the clergy joined hands with their traditional ally, the military, to plot the overthrow of his government. Bhutto's attempt to establish an authoritarian rule led him to rely more and more on the coercive apparatus of the state and the intelligence agencies.

Bhutto did little to strengthen democratic institutions and to make the process of democratic reform irreversible. Instead, his entire effort was aimed at promoting a personalized rule. He did not trust anyone. Given his overwhelming paranoia and insecurity,

Bhutto geared up the ISI to keep surveillance not only on his opponents, but also on his own party men and cabinet ministers. The agency kept dossiers on politicians, bureaucrats, judges and anyone else considered important. The collapse of democratic institutions and the Constitution's loss of sanctity created a vacuum of authority that provided a favourable condition for the Bonapartist generals. Bhutto's use of the army to crush the uprising in Pakistan's western province of Balochistan provided an opportunity to the military to reassert itself in the country's politics. Bhutto's politics of expediency and attempts to appease the country's religious lobby allowed the Islamists, who were routed in the 1971 elections, to revive themselves. A nationwide agitation, led by right wing Islamic parties in the aftermath of the controversial elections in 1977, shook his government. In a desperate attempt to appease the Islamists, Bhutto prohibited the sale and use of alcoholic beverages, banned gambling and closed down nightclubs. But the die had already been cast.

Bhutto had handpicked General Zia ul-Haq, a lesser-known officer, for the post of army chief in 1976 over the heads of half a dozen senior officers, believing he did not have any personal ambitions. General Zia was a devout Muslim and Bhutto thought he would never betray his trust. Once elevated to the top position, the General did not take much time to develop secret contacts with hardline religious groups and conspired with them to overthrow his benefactor. On 5 July 1977, a bloodless military coup brought an end to Bhutto's six-and-a-half-year civilian rule. There was a complete convergence of interests between the Islamists and the military leadership which had ousted the Bhutto government. Several accounts of Bhutto's last days in power revealed that Major-General Ghulam Jilani, Bhutto's handpicked chief of the ISI, played a key role in the coup plan. Two years later, in April 1979, Bhutto was hanged after a dubious trial on charges of murdering a political rival. A secularist, elected with an overwhelming mandate in 1970, Bhutto probably had the best chance of any leader of taking on the Islamists and winning. His inglorious demise can only fill Pakistan's current leader, attempting a similar task, with foreboding.

Bhutto's usurper, General Zia, came from a humble lowermiddle-class background. His father, a religious man, held a clerical government post during British rule in India. Despite his poverty, he provided his son with a decent education. Zia was educated at Delhi's most prestigious St Stephen's College before getting a commission in the Royal Indian Army in 1942. He was a captain at the time of the creation of Pakistan in 1947. He was a hard-working officer, but lacked brilliance. Many of his colleagues believed he was lucky to be promoted beyond Colonel. From a young age, Zia was actively involved with the popular religious movement, Tablighi Jamaat, 6 one of the most influential grassroots Islamic movements in the South Asian subcontinent, which had hardline traditionalist views on the role of Islam in modern society.

A short man with a thick moustache, Zia bore a strong resemblance to the British comedian, Terry-Thomas. But he was by no means a fool. He was shrewd enough to know how to make his way up. He convinced Bhutto of his absolute loyalty and more importantly of his incapacity to be otherwise. Zia was not only an authoritarian; he also aspired to turn Pakistan into an ideological state ruled by strict Islamic sharia laws. He was an unpopular and controversial leader, whose survival in power largely owed to the external factors that emerged after the invasion by Soviet forces of Pakistan's northern neighbour, Afghanistan, in 1979. Pakistan became a frontline state and a bulwark in the West's war against communism.

General Zia brought in his close confidant, Lt.-General Akhter Abdur Rehman, to head the ISI in 1979, as the spy agency became a crucial cog in the resistance against the Soviet forces. Its tentacles started to spread far and wide. An artillery officer, General Rehman 'had a cold, reserved personality, almost inscrutable, always secretive, with no intimates except his family.' His officers remembered him as a hard man to serve due to his brusque manner and reputation as a strict disciplinarian. 'That made him an ideal man for the job,' said Brigadier Mohammed Yousuf, a former ISI officer who served under him. General Rehman ran the organization until 1987. He was not a radical Islamist, but a staunch Muslim nationalist. Under General Rehman, the ISI, in partnership with the CIA, conducted the biggest covert operation in modern history. The two organizations had secretly collaborated for years, yet General Zia was not prepared to give a free hand to the CIA. He laid down strict rules to ensure that the ISI would maintain control over contacts with Afghan mujahidin groups. No CIA operatives would be allowed to cross the border into Afghanistan. Distribution of weapons to the Afghan commanders would be handled only by the ISI. All training of mujahidin would be the sole responsibility of the ISI. While the CIA supplied money and weapons, it was the ISI that moved them into Afghanistan. The Americans relied almost entirely on the Pakistani service to allocate weapons to the mujahidin groups. The framework provided the Pakistani spy agency with almost total control over the covert operation inside Afghanistan. The decade-long secret war raised the organization's profile and gave it huge clout. With the active help of the CIA and Saudi Arabia, the ISI turned the Afghan resistance into Islam's holy war.

The decade-long conflict in Afghanistan gave the Islamic extremists a rallying point and a training field. Young Muslims around the world flocked to Afghanistan to fight against a foreign invader. Some 35,000 holy warriors joined the Afghan war from 1979 to 1989. The largest number came from the Middle East. Some were Saudis, among whom was a young man called Osama bin Laden.

The 23-year-old son of a wealthy construction magnate said he arrived in Peshawar for the first time in 1980. He was a very shy person who had no experience of war and, as a fresh graduate from the King Abdul Aziz University, bin Laden had left the luxury of his home to help the mujahidin fighting jihad in Afghanistan. It was the beginning of his long association with Afghan and Pakistani Islamic radicals. By 1984 he was spending most of his time in Peshawar, sometimes with Abdullah Azzam, a Palestinian Islamic scholar and an ideologue who inspired the rush of foreign Muslims to fight in Afghanistan. It was also during this period that he developed links with the ISI, which had become the main conduit for arms and training for the Islamic warriors. The Afghan war placed enormous resources at the agency's disposal. Weapons provided by the CIA were channelled to Afghan fighters through the ISI. By the mid 1980s, every dollar given by the CIA was matched by another from Saudi Arabia. The funds, running into several million dollars a year, were transferred by the CIA to the ISI's special accounts in Pakistan. The backing of the CIA and the funnelling of massive amounts of US military aid helped the ISI develop into 'a parallel structure wielding enormous power over all aspects of government.' It became all-pervasive and there was an unprecedented expansion in the surveillance of political opponents.

Even those in the army were not spared. According to a former senior intelligence officer, part of the ISI's function was to keep a careful watch on the generals and ensure their loyalty to the regime. In the words of Brigadier Yousaf, 'So powerful had the ISI become that apprehension, even fear, of what the ISI could do, was real.' The Afghan war provided the agency with the public profile that it did not have before. Former ISI officials point out that the general expansion the agency underwent as a consequence of its Afghan role had a bearing on its overall functioning. While its role in supervising a successful covert operation in Afghanistan won the ISI international acclaim, at home it was a dreaded name.

The Afghan resistance was projected as part of the global jihad against communism. The ISI's training of guerrillas was integrated with the teaching of Islam. The prominent theme was that Islam was a complete socio-political ideology under threat from atheistic communists. The Afghan war produced a new radical Islamic movement. Besides the holy warriors from Islamic countries, thousands more joined madrasas set up with funding from Saudi Arabia and some other Muslim countries. More than 100,000 foreign Muslim radicals were directly influenced by the Afghan jihad. The military's involvement in the politics of religion, however, had started long before the Afghan jihad. General Zia not only ushered Pakistan into its longest period of military rule, but also tried to turn the country into an ideological state. He extended the army's role in domestic politics much further than defined by earlier military rulers. Previously, the military was seen as the ultimate guarantor of the country's territorial integrity and internal security.

But Zia expanded its role as the defender of Pakistan's ideological frontiers as well.

Addressing army officers, Zia argued that, as Pakistan was created on the basis of the two-nation theory and Islamic ideology, it was the duty of the 'soldiers of Islam to safeguard its security, integrity and sovereignty at all costs, both from internal turmoil and external aggression'. He claimed the state was created exclusively to provide its people with the opportunity to follow 'the Islamic way of life.' Preservation of the country's Islamic character was seen to be as important as the security of the country's geographical frontiers. Zia's military rule differed by its very serious attempt to create an ideological basis for all areas of activities of the state and society. General Zia cleverly used Islam to consolidate his power and legitimize his military rule. The Islamization process became the most important feature of his eleven- year rule and its raison d'e^tre. For General Zia, Pakistan was an Islamic ideological state and therefore politics could not be separated from religion. He believed that he had a divine obligation to establish an Islamic society ruled in accordance with the Qur'an and sharia. Afraid to face a free electorate and having no mandate to govern, the General turned to Allah. He introduced a rigid interpretation of Islamic sharia, thus empowering the clergy. The move to Islamize the state found ready allies among the religious parties, many of which already had close ties with the military. The Jamaat-i-Islami and other Islamic groups were co-opted by the government. For the first time in Pakistan's history, the Islamists occupied important government positions. Being in power helped Jamaat-i-Islami, the most influential and organized mainstream religious party, penetrate state institutions. Thousands of party activists and sympathizers were given jobs in the judiciary, the civil service and educational institutions. These appointments strengthened the hold of the Islamists on crucial state apparatuses for many years to come.

The process of the Islamization of the state and society took place on two levels. Firstly, changes were instituted in the legal system. Sharia courts were established to try cases under Islamic laws. For the first time, the government assumed the role of the collector of religious taxes. Secondly, Islamization was promoted through the print media, radio, television and mosques. Through a series of religious decrees, the government moved to Islamize the civil service, the armed forces and the education system. School textbooks were overhauled to ensure their ideological purity. Books deemed un-Islamic were removed from syllabuses and university libraries. It was made compulsory for civil servants to pray five times a day. Confidential reports of government officials included a section in which the staff were given marks for regularly attending prayers and for having a good knowledge of Islam. General Zia's move to Islamize the army carried the most critical implications. Pakistan's army, carved out of the British Indian army, inherited British traditions and remained a secular organization until General Zia tried to give it a new Islamic orientation. Islam was incorporated into the army's organizational fabric. For the first time Islamic teachings were introduced into the Pakistan Military Academy.

Islamic training and philosophy were made a part of the curriculum at the Command and Staff College. A Directorate of Religious Instruction was instituted to educate the officer corps on Islam. Islamic education also became a part of the promotion exams. The officers were required to read The Quranic Concept of War, a book written by a serving officer, Brigadier S. K. Malik. The officers were taught to be not just professional soldiers, but also soldiers of Islam.

In his foreword to the book, General Zia wrote: 'The professional soldier in a Muslim army, pursuing the goals of a Muslim State, cannot become "professional" if in all his activities he does not take on "the colour of Allah".' To gain promotion an officer was required to be a devout Muslim. Scores of highly professional and secular officers were sidelined for not meeting the criterion of a 'good Muslim'. As a consequence of this policy, many conservative officers reached the senior command level. Radical Islamist ideology permeated the army with the free flow of religious political literature in the armed forces training institutions. Friday prayers at regimental mosques, a matter of individual choice in the past, became obligatory. Mullahs belonging to the Deobandi sect were appointed to work among the troops.

They were supposed to be the bridge for officers between the westernized profession and the faith. Units were required to take noncombatant mullahs with them to the front line. Soldiers were encouraged to attend 'Tablighi' gatherings. The purpose was to indoctrinate cadets and young officers with an obscurantist interpretation of Islam. Many of those cadets later rose to positions of power and took control of sensitive institutions, including the ISI. The change in the social composition of the officer corps also led to the ideological reorientation of Pakistan's army in the 1980s. The army officer corps, till the 1970s, was elitist in character and composition.

But the trend changed. In the past, members of the upper class or the rural aristocracy had largely dominated the army ranks; now the new officers mostly came from the lower and middle social strata. The rank and file of junior officers since the 1970s had come, not from prosperous central Punjab, but from the impoverished and relatively backward northern districts where a fundamentalist religious ethos prevailed. Unlike their predecessors, the new officers did not come from elite English-language schools, but were the product of modest state-run educational institutions. The spirit of liberalism, common in the 'old' army, was practically unknown to them. They were products of a social class that, by its very nature, was conservative and easily influenced by Islamic fundamentalism. They were distinctively less westernized.

Their narrow nationalist orientation and education, professional training and culture made them very receptive to the influence of Islamic groups, particularly to the Jamaat-i-Islami (JI), which freely carried out propaganda work among the officers and circulated Islamic literature. Most importantly, the JI's propaganda among vast numbers of troops was officially sanctioned by commanding officers on the battalion level and above. That led to the evolution of a long-lasting alliance between the military and the mosque. The main objective of the JI was to penetrate the army and use it to seize state power. The practice introduced by General Zia of sending combat officers to universities in Pakistan, in which the JI often had pervasive influence, facilitated the party's objective. The influence of the JI was particularly visible among officers from rural and semi-urban backgrounds. They had a high level of religious intolerance and held extremist views regarding 'external enemies' and the threat to Islam.

The number of officers sporting beards for religious reasons had visibly risen. These young officers constituted the main base for General Zia's regime. According to a retired general, 25 to 30 per cent of the officers had Islamic fundamentalist leanings. Many of them regularly attended 'Tablighi' groups that propagated the faith. They would take time off to join missionary bands preaching a return to purist Islamic values and recruiting other Muslim men to join them. Thousands of soldiers and army officers would join the annual gathering of the TJ at Raiwind on the outskirts of Lahore. More than one million faithful from across the country and abroad attend this Muslim congregation, the largest after the Haj in Mecca. Even the liberals adopted the philosophy on issues like jihad and support for militancy in Kashmir. The humiliating military defeat by the Indian army in the 1971 war, coupled with the disintegration of the country, also had a very deep impact on the psyche of army officers. Many young officers turned to religion for solace and became born-again Muslims. Given this situation, it was not surprising that radical Islamic officers made unsuccessful coup attempts in the mid 1980s. They wanted to bring about an Islamic revolution and establish a theocratic state. The fact that the attempts were crushed and the rebel officers were charged with treason indicated that, notwithstanding its Islamization, the state was not prepared to tolerate a theocratic rule.

The situation in Afghanistan provided inspiration to a whole generation of Pakistani Islamic radicals who considered it their religious duty to fight the oppression of Muslims anywhere in the world. It gave a new dimension to the idea of jihad, which till then had only been employed by the Pakistani state in the context of mobilizing the population against the arch rival – India. The Afghan war saw the privatization of the concept of jihad. Militant groups emerged from the ranks of traditional religious movements, who took the path of an armed struggle for the cause of Islam. The ISI's active role in the Afghan jihad brought Pakistani army officers into direct contact with the radical Islamists.

The handling of jihad also indoctrinated the military and intelligence officers. At least two former ISI chiefs – General Hamid Gul and General Javed Nasir – remained actively involved with Islamic radical movements. Both promoted pan-Islamism and strove for an Islamic revolution that would free Pakistan from perceived western, and particularly American, cultural and political influences. General Gul, who liked to call himself a 'Muslim visionary', succeeded General Rehman in 1987 when the Afghan resistance had entered its most critical juncture.

The most charismatic of the ISI chiefs, he openly aligned himself with the hardline Afghan mujahidin leader, Gulbuddin Hekmatyar, and right-wing Pakistani Islamic parties. He would later take credit for dismantling the Soviet Union. General Gul had worked closely with the CIA, but turned anti-American after the Geneva Accord in 1987, which paved the way for the withdrawal of Soviet forces from Afghanistan. The Accord was signed by the Pakistani civilian government led by Prime Minister Mohammed Khan Junejo against the wishes of President Zia, who believed it deprived the Afghan mujahidin of the opportunity to take over Kabul. Resentment among the military and ISI officers deepened in 1990, just one year after the Soviets pulled out of Afghanistan and the US administration imposed sanctions on Pakistan for its nuclear programme. The CIA ended its support for the Afghan rebels and its links with the ISI went cold. It was a sudden end to the strong link between the two agencies. The military leadership accused the United States of dumping Pakistan after the collapse of the Soviet Union. Long before the Afghan war was over, General Gul had started organizing a new jihad front in Kashmir. 'It is the years of our work that has realized the armed uprising against the Indian forces in Kashmir,' he told me in early 1990 when he was commanding the country's elite armoured corps.

The death of General Zia ul-Haq in a mysterious air crash in August 1988 brought an end to Pakistan's longest-serving military government. It left the generals with the choice of either imposing martial law again or holding elections and transferring power to a democratically elected civilian government. They went for the second option, realizing that a perpetuation of military rule might provoke public resistance and exacerbate the turmoil in an already highly polarised society. However, the generals were not prepared to pull out completely and leave the political field solely to the politicians, particularly to the daughter of Zulfikar Ali Bhutto, Benazir Bhutto, who now led the Pakistan People's Party, which was certain to sweep the polls. To contain Benazir, the military cobbled together the Islamic Democratic Alliance (IDA), uniting all the right-wing parties under the leadership of Nawaz Sharif. General Aslam Beg, then Chief of Army Staff, and General Gul justified the move by saying it was necessary for the viable functioning of a democratic system.

The main objective, however, was to ensure the military's continued role in the new system, overseen by the generals. By creating a counterbalance in the form of the IDA, the generals constrained the new government of a political party that had led the resistance against military hegemony for ten years. The military-sponsored alliance comprised a mix of traditional power brokers, religious parties and politicians who had emerged on the scene during the 1980s under military patronage. General Zia's regime, needing a measure of legitimacy and a social base of support, had co-opted segments of Punjab's dominant socio-economic strata: influential landlords, industrialists and emerging commercial groups. The new leadership of the Pakistan Muslim League thus largely consisted of politicians who owed their political rise to the military's backing. The other important components of the alliance were the right-wing Islamic groups, including the Jamaat-i-Islami, which had been involved in the Afghan jihad against the Soviet forces and the separatist war in Indian-controlled Kashmir.

While the IDA, which carried General Zia's legacy, could not achieve any significant electoral gains in the 1988 elections in the three smaller provinces, it did relatively well in Punjab and prevented the PPP from winning an absolute majority in the National Assembly. The military reluctantly handed over power to Benazir at the centre, but prevented her party from forming the government in Punjab, the biggest and most important of Pakistan's four provinces, which was being ruled by Nawaz Sharif. Political manipulation by the ISI helped Sharif retain power in the province with the support of independent members. Benazir Bhutto's assumption of power, touted at the time as the dawn of a new democratic era, was in fact a transition from direct to indirect military rule. The formal restoration of civilian rule in 1988 did not reduce the ISI's clout. A return to the barracks did not mean that the military's structure of control and manipulation had been dismantled. The ISI still kept a close watch on civilian rulers. The strains in civil and military relationships remained the biggest obstacle to democracy taking root in Pakistan.

The Chief of Army Staff remained the power behind the scenes in alliance with the new President, Ghulam Ishaq Khan, who held sweeping powers under the Eighth Amendment in the constitution introduced by General Zia. For the army, the new political situation offered power without responsibility. The military continued to oversee Pakistan's policies on Afghanistan and India, managing relations with the USA and controlling the country's nuclear weapons programme. The army high command's decision to rest content with dominance, rather than direct intervention, was based on a careful calculation of the advantages and disadvantages of playing arbiter in Pakistan's highly polarized and conflict-ridden political scene. A hamstrung Benazir Bhutto found herself directly clashing with the army when she removed General Gul from the ISI. The General, who was the architect of the IDA, had never reconciled himself to even restricted civilian rule. Benazir was never trusted by the establishment and many generals made it a point not to salute her. It was in February 1990 when I met General Gul in Multan, where he was posted as the commander of Pakistan's main strike corps, after being sacked from the ISI. It was the period when the army had just concluded a major war exercise called 'Zarb-i-Momin'. The General accused Prime Minister Benazir of trying to make peace with India and questioned her patriotism. 'By conducting the exercise we have blocked her designs to undermine Pakistan's defence,' he said.

It was apparent that the generals were looking for an opportune moment to remove Benazir from power. The clash came to a head when the Prime Minister tried to retire the Chairman of the Joint Chiefs of Staff Committee, Admiral Iftikhar Ahmed Sarohi, and gave a year's extension in service to Lt.-General Alam Jan Mehsud, a senior general believed to be close to her government. The army top brass saw this as blatant interference in the army's internal affairs. Eventually, on 6 August 1990, President Ghulam Ishaq Khan dismissed her government on charges of corruption and dissolved the National Assembly. President Khan, a former top bureaucrat, was the central pillar of General Zia's military rule and represented his legacy. Benazir's party met a humiliating defeat in the following elections, losing the government even in her home province and political stronghold, Sindh.

The military leadership did not want to take any chances that time. To ensure the IDA's electoral victory, the ISI financed the election campaign of many top Alliance leaders. The list, which was published in national newspapers, contained the names of some leading politicians, including Nawaz Sharif. Most of them later held important posts in the new government. In 1988, the long-simmering political discontent in Kashmir exploded into a popular uprising against Indian control. Thousands of young Kashmiris crossed over to the Pakistani side of the Line of Control to receive guerrilla training as the Indian authorities tried to crush the insurgency by brute force. Essentially an indigenous rebellion, it soon turned into an armed struggle against Indian forces. Thousands of Kashmiri youths joined the separatist struggle, which was initially led by the independent nationalist organization, Jammu and Kashmir Liberation Front (JKLF)

The ISI used the extensive intelligence and militant network that it had built up during the Afghan war to support a new jihad against the Indian forces in Kashmir. In Afghanistan in the 1980s, jihadist cadres came from the ranks of motivated Islamists across the Muslim world who were prepared to die for the cause, as well as kill the 'communists'. The spirit saw its continuation in Kashmir, which became one of the world's hottest 'Islamic jihad' spots. The ISI's involvement increased further in the early 1990s when, in an attempt to sideline the JKLF, it started supporting the pro-Pakistani Hezb-ul Mujahideen. The move divided the Kashmiri struggle and led to internecine battles. The mid 1990s saw the increasing role of Pakistani-based militant groups. That was when Harkat-ul-Mujahideen (HuM), Lashkar-e-Taiba (LeT) and later Jaish-e-Mohammed (JeM) emerged as the main guerrilla forces in the disputed state. Those hardline Islamic groups changed the complexion of the struggle.

SUPPORTING TERRORISM?

"The ISI probably would not define what they've done in the past as 'terrorism,'" says William Milam, former U.S. ambassador to Pakistan. Nevertheless, experts say the ISI has supported a number of militant groups in the disputed Kashmir region between Pakistan and India, some of which are on the U.S. State Department's Foreign Terrorist Organizations list.

While Pakistan has a formidable military presence near the Indian border, some experts believe the relationship between the military and some Kashmiri groups has greatly changed with the rise of militancy within Pakistan. Shuja Nawaz, author of*Crossed Swords: Pakistan, its Army, and the Wars Within,* says the ISI "has certainly lost control" of Kashmiri militant groups. According to Nawaz, some of the groups trained by the ISI to fuel insurgency in Kashmir have been implicated in bombings and attacks within Pakistan, therefore making them army targets.

"I do not accept the thesis that the ISI is a rogue organization." —William Milam, former U.S. ambassador to Pakistan.

On Pakistan's western border with Afghanistan, the ISI supported the Taliban up to September 11, 2001, though Pakistani officials deny any current support for the group.

Pakistan's government was also one of three countries, along with the United Arab Emirates and Saudi Arabia, that recognized the Taliban government in Afghanistan. The ISI's first major involvement in Afghanistan came after the Soviet invasion in 1979, when it partnered with the CIA to provide weapons, money, intelligence, and training to the mujahadeen fighting the Red Army.

At the time, some voices within the United States questioned the degree to which Pakistani intelligence favoured extremist and anti-American fighters. Following the Soviet withdrawal, the ISI continued its involvement in Afghanistan, first supporting resistance fighters opposed to Moscow's puppet government, and later the Taliban.

Pakistan stands accused of allowing that support to continue. Afghan President Hamid Karzai has repeatedly said Pakistan trains militants and sends them across the border. In May 2006, the British chief of staff for southern Afghanistan told the *Guardian*, "The thinking piece of the Taliban is out of Quetta in Pakistan. It's the major headquarters." Speaking at the Council on Foreign Relations in September 2006, then-president Pervez Musharraf responded to such accusations, saying, "It is the most ridiculous thought that the Taliban headquarters can be in Quetta." Nevertheless, experts generally suspect Pakistan still provides some support to the Taliban, though probably not to the extent it did in the past. "If they're giving them support, it's access back and forth [to Afghanistan] and the ability to find safe haven," says Kathy Gannon, who covered the region for decades for the Associated Press. Gannon adds that the Afghan Taliban needs Pakistan even less as a safe haven now "because [it has] gained control of more territory inside Afghanistan."

Many in the Pakistani government, including slain former prime minister Benazir Bhutto, have called the intelligence agency "a state within a state," working beyond the government's control and pursuing its own foreign policy. But Nawaz says the intelligence agency does not function independently. "It aligns itself to the power center," and does what the government or the army asks it to do, says Nawaz.

Control over the ISI

Constitutionally, the agency is accountable to the prime minister, says Hassan Abbas, research fellow at Harvard's Kennedy School of Government. But most officers in the ISI are from the army, so that is where their loyalties and interests lie, he says. Experts say until the end of 2007, as army chief and president, Musharraf exercised firm control over the intelligence agency. But experts say it is not clear how much control Pakistan's civilian government—led by Bhutto's widower, President Asif Ali Zardari—

has over the agency. In July 2008, the Pakistani government announced the ISI will be brought under the control of the interior ministry, but revoked its decision (BBC) within hours. Bruce Riedel, an expert on South Asia at the Brookings Institution, says the civilian leadership has "virtually no control" (PDF) over the army and the ISI. In September 2008, army chief Ashfaq Parvez Kiyani replaced the ISI chief picked by former president Musharraf with Lt. Gen. Ahmed Shuja Pasha. Until then, Pasha headed military operations against militants in the tribal areas. Some experts said the move signaled that Kiyani was consolidating his control over the intelligence agency by appointing his man at the top. In November 2008, the government disbanded ISI's political wing, which politicians say was responsible for interfering in domestic politics. Some experts saw it as a move by the army, which faced much criticism when Musharraf was at the helm, to distance itself from politics.

"I do not accept the thesis that the ISI is a rogue organization," Milam says. "It's a disciplined army unit that does what it's told, though it may push the envelope sometimes." With a reported staff of ten thousand, ISI is hardly monolithic: "Like in any secret service, there are rogue elements," says Frederic Grare, a South Asia expert and visiting scholar at the Carnegie Endowment for International Peace. He points out that many of the ISI's agents have ethnic and cultural ties to Afghan insurgents, and naturally sympathize with them. Marvin G. Weinbaum, an expert on Afghanistan and Pakistan at the Middle East Institute, says Pakistan has sent "retired" ISI agents on missions the government could not officially endorse.

Resistance in FATA

Pakistan's tribal areas along the Afghan border have emerged as safe havens for terrorists. Experts say because of their links to the Taliban and other militant groups, the ISI has some influence in the region. But with the mushrooming of armed groups in the tribal agencies, it is hard to say which ones the agency controls. Also, there appears to be divisions within the ISI. While some within the intelligence agency continue to sympathize with the militant groups, Harvard's Abbas says others realize they cannot follow a policy contradictory to that of the army, which is directly involved in counterterrorism operations in the area.

Bruce Riedel, an expert on South Asia at the Brookings Institution, says the civilian leadership has "virtually no control" over the army and the ISI.

Mixed Record on Counterterrorism

Pakistan has arrested scores of al-Qaeda affiliates, including Khalid Sheikh Mohammed, the alleged mastermind of the 9/11 attacks. The ISI and the Pakistani military have worked effectively with the United States to pursue the remnants of al-Qaeda. Following 9/11, Pakistan also stationed eighty thousand troops in the troubled province of Waziristan near the Afghan border. Hundreds of Pakistani soldiers died there in resulting clashes with militants, which, as Musharraf told a CFR meeting in September 2006, "broke the al-Qaeda network's back in Pakistan."

But Musharraf did crack down on terrorist groups selectively, as this Backgrounder points out. Weinbaum in 2006 said the Pakistani military has largely ignored Taliban fighters on its soil. "There are extremist groups that are beyond the pale with which the ISI has no influence at all," he says. "Those are the ones they go after." In 2008, Ashley J. Tellis, senior associate at the Carnegie Endowment for International Peace, wrote (PDF) in *TheWashington Quarterly* that Musharraf tightened pressure on groups whose objectives were out of sync with the military's perception of Pakistan's national interest.

The Taliban as a Strategic Asset

Pakistan does not enjoy good relations with the current leadership of Afghanistan, partly because of rhetorical clashes with Afghan President Hamid Karzai, and partly because Karzai has forged strong ties with India. But there have been increased efforts by the United States to close this gap. The Obama administration's regional strategy unveiled in March 2009 focused on creating new diplomatic mechanisms; a trilateral summit of the leaders of the United States, Pakistan, and Afghanistan has been one such step toward helping reduce the level of distrust that runs among all three countries. But lingering suspicions about ISI's support for the Taliban continue to pose problems. In an October 2006 interview, Musharraf said some retired ISI operatives could be abetting the Taliban insurgency in Afghanistan, but he denied any active links. Zardari, too, denies any ISI links with the Taliban or al-Qaeda. In a May 2009 interview with CNN, he remarked all intelligence agencies have their sources in militant organizations but that does not translate to support. "Does that mean CIA has direct links with al-Qaeda? No, they have their sources. We have our sources. Everybody has sources."

Some experts say Pakistan wants to see a stable, friendlier government emerge in Afghanistan. Though the insurgency certainly doesn't serve this goal; increased Taliban influence, especially in the government, might. Supporting the Taliban also allows Pakistan to hedge its bets should the NATO coalition pull out of Afghanistan. In a February 2008interview with CFR.org, Tellis said the Pakistani intelligence services continue to support the Taliban because they see the Taliban leadership "as a strategic asset," a reliable back-up force in case things go sour in Afghanistan.

Not everyone agrees with this analysis. According to Weinbaum, Pakistan has two policies. One is an official policy of promoting stability in Afghanistan; the other is an unofficial policy of supporting jihadis in order to appease political forces within Pakistan. "The second [policy] undermines the first one," he says. Nawaz says there is ambivalence within the army regarding support for the Taliban. "They'd rather not deal with the Afghan Taliban as an adversary," he says.

ALLEGATIONS OF TERRORIST ATTACKS

Indian officials implicated the ISI for the November 2008 terrorist attacks in Mumbai that killed nearly two hundred people. India's foreign ministry said the ISI had links (Reuters)to the planners of the attacks, the banned militant group Lashkar-e-Taiba, which New Delhi blames for the assault. Islamabad denies allegations of any official involvement, but acknowledged in February 2009 that the attack was launched and partly planned (AP)from Pakistan. The Pakistani government has also detained several Islamist leaders, some of them named by India as planners of the Mumbai assault. Gannon says this is an unusual step by Pakistan, which never got enough credit in India because the country was in the middle of a national election. "I don't see any evidence" to believe that the ISI was behind the Mumbai attack, she says. However, she doubts the agency has severed all its ties with groups like Lashkar-e-Taiba which it supported to fight in Indian-administered Kashmir.Indian officials also claim to have evidence that the ISI planned the July 2006 bombing of the Mumbai commuter trains, but these charges seem unlikely to some observers of the long, difficult India-Pakistan relationship. The two nations have a history of finger-pointing, and while some of the allegations hold water, there is a tendency to exaggerate.

Following the release of the British report regarding its July 7, 2005, bombings of London's mass transit system—which London insists is not a statement of policy—Weinbaum said it makes "too broad a statement." Though Pakistan does offer safe haven to Kashmiri groups, and perhaps some Taliban fighters, the suggestion that the ISI is responsible for the 7/7 bombings is "a real stretch," Gannon says.

THE PAKISTANI INTELLIGENCE AGENCY (ISI) AND TERRORISM

Pakistan's intelligence agency, the Inter-Services Intelligence (ISI), has long faced accusations of meddling in the affairs of its neighbors. A range of officials inside and outside Pakistan have stepped up suggestions of links between the ISI and terrorist groups in recent years. In autumn 2006, a leaked report by a British Defense Ministry think tank charged, "Indirectly Pakistan (through the ISI) has been supporting terrorism and extremism—whether in London on 7/7, or in Afghanistan, or Iraq." In India, Mumbai's police chief has claimed to have proof that the ISI planned the July 2006 bombing of the Indian city's commuter rail system, which was carried out by the Kashmir-based militant group Lashkar-e-Taiba. After massive bombings marred former Prime Minister Benazir Bhutto's October 2007 return to Pakistan, some of her supporters blamed the ISI; Bhutto herself called for the sacking of the country's intelligence chief.

In a September 2006 BBC interview, President Pervez Musharraf underscored the importance of his nation's role: "Remember my words: if the ISI is not with you and Pakistan is not with you, you will lose in Afghanistan." In a later appearance on NBC's "Meet the Press," Musharraf acknowledged some retired ISI operatives could be abetting the Taliban insurgency in Afghanistan. But Pakistan's government denies allegations of supporting terrorism, citing as evidence its cooperation in the "war on terror," in which it has taken significant losses, politically and on the battlefield.

Does ISI Support Terrorists?

"The ISI probably would not define what they've done in the past as 'terrorism,'" says William Milam, former U.S. ambassador to Pakistan. Nevertheless, experts say the ISI has supported a number of militant groups in the disputed Kashmir region between Pakistan and India, some of which are on the State Department's Foreign Terrorist Organizations list. Though the level of assistance to these groups has varied, Kathy Gannon, who covered the region for decades for the Associated Press, says previous support consisted of money, weapons, and training.

Though Pakistani officials deny any current support for the Taliban—which the State Department does not deem a terrorist group—the ISI certainly has supported Afghan insurgents in the past. ISI's first major involvement in Afghanistan came after the Soviet invasion in 1979, when it partnered with the CIA to provide weapons, money, intelligence, and training to the mujahadeen fighting the Red Army. At the time, some voices within the United States questioned the degree to which Pakistani intelligence favoured extremist and anti-American fighters. Following the Soviet withdrawal, the ISI continued its involvement in Afghanistan, first supporting resistance fighters opposed to Moscow's puppet government, and later the Taliban.

Pakistan stands accused of allowing that support to continue. Afghan President Hamid Karzai has repeatedly said Pakistan trains militants and sends them across the border. In May 2006, the British chief of staff for southern Afghanistan told the Guardian, "The thinking piece of the Taliban is out of Quetta in Pakistan. It's the major headquarters." Speaking at the Council on Foreign Relations in September, President Musharraf responded to such accusations, saying, "It is the most ridiculous thought that the Taliban headquarters can be in Quetta." Nevertheless, experts generally suspect Pakistan still provides some support to the Taliban, though probably not to the extent it did in the past. "If they're giving them support," Gannon says, "it's access back and forth [to Afghanistan] and the ability to find safe haven."

How much Control does Pakistan's political Leadership have over the ISI?

Experts say Musharraf exercises firm control over his intelligence agency. "I do not accept the thesis that the ISI is a rogue organization," Milam says. "It's a disciplined army unit that does what it's told, though it may push the envelope sometimes." With a reported staff of 10,000, ISI is hardly monolithic: "Like in any secret service, there are rogue elements," says Frederic Grare, a South Asia expert and visiting scholar at the Carnegie Endowment. He points out that many of the ISI's agents have ethnic and cultural ties to Afghan insurgents, and naturally sympathize with them.

President Musharraf's admission that retired ISI agents may be helping Taliban fighters suggests his government knows of at least some unsanctioned Pakistani support for the Afghan insurgency. Experts note Musharraf's acknowledgement also gives him plausible deniability of any sanctioned assistance Pakistan may also be providing. Marvin G. Weinbaum, an expert on Afghanistan and Pakistan at the Middle East Institute, says Pakistan has sent "retired" ISI agents on missions the government could not officially endorse. Some observers believe Pakistan's duplicity is deliberate: "Musharraf's been playing with us since day one," Grare says.

What has Pakistan done to Combat Terrorism?

Pakistan has arrested scores of al-Qaeda affiliates, including Khalid Sheikh Mohammed, the alleged mastermind of the 9/11 attacks. The ISI and the Pakistani military have worked effectively with the United States to pursue the remnants of al-Qaeda. Pakistan also stationed 80,000 troops in the troubled province of Waziristan near the Afghan border. Hundreds of Pakistani soldiers died there in resulting clashes with militants, which Musharraf told a recent CFR meeting "broke the al-Qaeda network's back in Pakistan." Though Pakistan has effectively battled al-Qaeda, Weinbaum says it has largely ignored Taliban fighters on its soil. "There are extremist groups that are beyond the pale with which the ISI has no influence at all," he says. "Those are the ones they go after."

What is Pakistan's Interest in aiding the Taliban?

Pakistan does not enjoy good relations with the current leadership of Afghanistan, partly because of rhetorical clashes with Afghan President Hamid Karzai, and partly because Karzai has made strong ties to India. Some experts say Pakistan wants to see a stable, friendlier government emerge in Afghanistan. Though the raging insurgency certainly doesn't serve this goal, increased Taliban influence, especially in the government, might. Supporting the Taliban also allows Pakistan to hedge its bets should the NATO coalition pull out of Afghanistan. Not everyone agrees with this analysis. According to Weinbaum, Pakistan has two policies. One is an official policy of promoting stability in Afghanistan; the other is an unofficial policy of supporting jihadis in order to appease political forces within Pakistan. "The second [policy] undermines the first one," he says.

Pakistan's interest in aiding Kashmiri Militants

Throughout its existence, Pakistan has viewed India as its enemy. Though Islamabad has a real interest in fostering better relations with New Delhi, little love is lost between the nations. Some experts say the Kashmiri militants keep pressure on the Indian government, though others point out the militants' actions have sparked tense showdowns between the two nuclear-armed states in the past. While Pakistan has a formidable military presence near the Indian border, Gannon says the militant groups there serve as "their second line of defense and offence."

British and Indian Charges against Pakistan

Experts are skeptical. Indian officials claim to have evidence that the ISI planned the July bombing of the Mumbai commuter trains, but the charges seem unlikely to some observers of the long, difficult India-Pakistan relationship. The two nations have a history of finger-pointing, and while some of the allegations hold water, there is a tendency to exaggerate. Furthermore, endorsing an attack on India would undermine Musharraf's own policy. "Pakistan is genuinely trying to open up relations with India," Gannon says.

In Weinbaum's view, the British report—which London insists is not a statement of policy—makes "too broad a statement." Though Pakistan does offer safe haven to Kashmiri groups, and perhaps some Taliban fighters, the suggestion that the ISI is responsible for the 7/7 bombings of London's mass transit system is "a real stretch," Gannon says.

Pakistan's Relationship with the Tribal Leaders in its Northwestern Provinces

Despite President Musharraf's claims of success in Pakistan's three-year military incursion into Waziristan, the region remains problematic. Local tribesmen are the traditional power brokers, and over the last two decades, they have earned large sums of money offering safe haven to Taliban and foreign fighters. At the outset of the U.S.-led war in Afghanistan, many militants fled across the border into Waziristan, putting pressure on the Pakistani government to address the situation. ISI officials met resistance when they asked local leaders to hand over foreign militants. Pakistan deployed troops to the area, though reports suggest that they took as many casualties as their adversaries. Unable to make military progress, Pakistani officials negotiated a peace deal with the local tribesmen, which was seen by many as a capitualtion to tribal influence in the area.

4

War on Terror: Intelligence 'Double Game'

Key US intelligence and strategic asset Central Intelligence Agency (CIA) and powerful Pakistani intelligence agency Inter-Services Intelligence (ISI) have hindered relations since the "War on Terror" begun in Afghanistan. Despite the soured relationship, CIA and ISI have worked in close cooperation with each other in certain issues. However, they do neither trust each other nor they are interested in it. Both the intelligence agencies have been working on the basis of their own calculation and interest rather than mutual interest. CIA has at times come under intense criticism for working with the ISI. The ISI has also been alleged and criticized for its role in Afghan war for protecting high-ranking militant figures. CIA believes that ISI supported targets like Osama bin Laden and Mullah Omar by providing them shelter. Both intelligence services seem to be playing a "double game," even as Pakistan joined US as an ally against the "war on terror," in post-9/11 scenario.

CIA no longer believes that ISI has been working in tandem in its effort to counter terrorism. ISI has been alleged of supporting terrorism and extremism even if they trace them. CIA station in Islamabad has a significant number of agents across the Pakistani soil. Both the CIA and ISI are eye and ear of their government policy and both have different doctrines and strategic goals in Afghanistan. They maintain an uneasy cooperation despite their common interest but they try to make an upper hand on several issues. US recognized Pakistan as a major Non-NATO Ally but CIA and ISI are the most unreliable strategic partner of counter-terrorism efforts. Relations between both the intelligence agencies have been uncomfortable and suspicious. CIA has prioritized on covert operation to target against al-Qaida, TTP and other foreign militants from neighbouring Afghanistan who plot attacks against US forces. They have huge intelligence network in this region as well a long list of dangerous militants. US military cross-border raid in Pakistani soil is aimed at hitting the ISI proxy militant's force.

BACKGROUND OF CIA, ISI COOPERATION

Both CIA and ISI earlier shared similar effort to support the Afghan and foreign mujahideen to covert-fight against Soviet forces in Afghanistan. They not only provided strategic and intelligence support for Afghan militia but also provided training, arms and funding in the 1980s. CIA had an impressive support from ISI role during the Cold War period when Pakistan received extensive military and other assistance from USA. Their relation was bitter after Soviet withdrew from Afghanistan in 1989. ISI solely engaged in the Afghan Civil War in 1990s and backed the Taliban regime. During that period, CIA was focused in Europe and Middle East and paid little attention on Pakistan and Afghanistan.

After seeing potential threat from Afghanistan, CIA agents/ informer in Pakistan and Afghanistan were at an unsurpassed level. After al-Qaeda attacked U.S. embassies in Kenya and Tanzania on August 7, 1998, CIA was a realizable frontier as well foreign policy tool for the US government. Their role increased by folds than the Cold War period. CIA's bin Laden mission began to hunting mission after they came from Sudan to Afghanistan. CIA local agent had a face-to-face with Laden and his fundamentalist supporters several times during the Taliban regime. Intelligence working on 'Bin Laden mission' had a huge investment. After increasing instances of terrorist attacks, CIA's only concern was to get hold of Laden – forget what's happening in Afghanistan.

As retaliation measures against bombings in US embassies, US forces started targeting Laden hideouts, including remote places.

On August 20, 1998, about 75 cruise missiles were fired by the U.S. at four Afghan training camps in Afghanistan from Arabian Sea through Pakistan's airspace. CIA's station in Pakistan provided the principle target list. However, Laden had left some hours before the missiles hit. US missile targeted Al Farouq training camp, Muawai camp – run by Pakistani Harkat-ul-Mujahideen to train militants, training camps in the Jarawah, the possible radical leaders' meeting point.

In the changed scenario in Afghanistan, CIA official's role became more influential. The CIA station chief in Pakistan is more powerful than its Ambassador. CIA's Pakistani station operator agent involved in tracing Bin Laden, directly informed the location in Afghanistan and provided information to the headquarters at the Counterterrorist Center. CIA directly briefed President Bill Clinton about the missile attack on their base. However, the President did not give orders to the military to strike on the Laden hideout. Following the 9/11 attacks, US demanded Taliban to handover laden and close the training camp, which the Taliban ignored. US force then launched "Operation Enduring Freedom" on 7 October 2001 which led to the downfall of Taliban regime in Afghanistan. Thereafter, CIA unit hunting Osama bin Laden dramatically increased its presence and activities in Pakistan and Afghanistan. After the Battle of Tora Bora in December 2001, laden and senior al Qaeda leaders escaped from there.

CIA has analysed that Pakistani tribal region is becoming a global hideout base for jihadist militants. The U.S. covert operations have continued in this region through pay role tribal agents. They also use local tribal leaders to fight against Taliban and other foreign fighters. Several bloody fights between tribal militants and foreign militia have been reported in remote Afghan borders. After failing their initial strategy to fight with foreign militia, CIA used several ground agents, intelligence network to trace the militia movement. However, foreign fighters gained control over tribal areas. CIA then used new approach of war and intensified drone campaign. Pakistan has described those attacks as an infringement of its sovereignty. But several air strikes have not yet gained concrete result but led to militant revenge through suicide bombings in return.

The CIA has often blamed ISI for supporting or planning several strikes and bombings against the US and western interest in Afghanistan. ISI's concern in Afghanistan has been different with that of USA. After potential strategic threat in Afghanistan, ISI joined hands with some proxy militant groups to secure their interest toll. But CIA easily traced the real channel and mentor of such hostile groups targeting the US and it's ally's interests.

Extremists' ties with ISI had alarmed the CIA. The US had repeatedly pressurized the Pakistani government to hunt down the militants. CIA has also handed over the list of Pakistani madrassas, where jihadists are fed against the US force in Afghanistan. Pakistan started several offensive operations in the tribal areas where foreign and local militia have set their safe haven. Operation was limited and eventually concluded with a peace deal. According to a US State Department data, since 2001 Pakistan has provided assistance in counter-terrorism efforts by capturing more than 600 al-Qaida members and their allies.

CIA not only faced the al-Qaida, their number one enemy, but also faced the most sophisticated insurgent group Haqqani network. The US has accused the ISI that this group has been established as its proxy force to serve Pakistani hidden interests in Afghanistan. US intelligence also accuses the Pakistani secret service of operating the Lashkar-e-Taiba to use the army's covert foreign mission. They still believe that ISI has not still abandoned to support militia. U.S. has accused Pakistan of using militants and killing of NATO troops in Afghanistan, which Pakistan denies. The US doesn't trust Pakistan's commitment to counter-terrorism efforts. Therefore, it raided the hideout of al-Qaeda chief Osama bin Laden in May 2011 and killed him. It also carries out drone strikes in the tribal area without informing the Pakistani security. According to New American Foundation data, 58 al-Qaeda and Taliban affiliated senior leaders have been killed in the CIA drone campaign in Pakistan from Afghanistan.

Who Gained the Ground?

Question remains, who achieved the strategic goal in Afghanistan? There are several players – not only USA, European

Union and Pakistan but Iran, India, Saudi Arabia, and UAE to Qatar have been playing the different roles. Afghanistan's central command has been in limbo after post 2014 because of foreign intelligence agencies' competition to get hold of the geopolitical ground.

Iran has also set of cultural, religious, political, and security relations with Afghanistan, which can be a ground to increase its influence. Tajik and Shi'a groups support Iran and are opposed to Taliban. Anti Iranian Baluchi insurgency group killed Iranian security personnel and Iran has security concern as well as Narcotics trafficking. Iran too has a strategic objective of defeating both US, Pakistan and Taliban because they do not want any other hostile force gain the ground in Afghanistan.

Islamic Revolutionary Guard Corps foreign wing 'The Quds Force' has been moving silently to the ground to secure their strategy. They have initiated an extensive covert campaign to play a key role to protect their nation's interest as well extensive influence over Afghan spies and paid informers' network. They want to seek a greater role after post 2014 Afghanistan like post US military present Iraq. Iran has certain concerns in Afghanistan because of it's geographical as well geopolitical situation.

India too has its own national security calculations, and has significantly increased its foreign intelligence agency Research and Analysis Wing (RAW) as well as its Military Intelligence (MI). Both RAW and MI have extensive influence in the government as well as in the security. India also increased its political and economic influence and launched different huge projects in Afghanistan. ISI fears Indian growing influence in Afghanistan and counter-measure strategy. New Delhi and Tehran has had close ties since the 1990s and both might continue to coordinate in the post 2014 Afghanistan. Behind the scene, both countries are actively engaged in countering Pakistani influence over Afghanistan and protect their joint interest. After concluding the fact, ISI played a double game with USA and adapted new tactics to gain the Afghan ground.

After realizing the loss of ground, ISI wants much closer relationship with CIA. On of the most recent examples are the recent prisons swap between Taliban and US government. But current overt action and open rhetoric cannot be the judgment of the final reality. This may take some more time unravel. Tehran's growing influence in Afghanistan is not favourable for its regional rival Saudi Arabia. Saudi intelligence agency General Intelligence Presidency (GIP) is also playing a role in Afghanistan. Behind the scenes, GIP continues to engage with ISI and Taliban representatives and may have been directly in touch with Taliban figures. The British Secret Intelligence Service MI6 idea is to hold talks with moderate Taliban elements and end the conflict. Other EU intelligence agencies consent with the British. All the intelligence agencies believe that fugitive senior Afghan Taliban leaders are safe in ISI protection.

How Pakistan Deals its Internal Conflict?

USA insisted Pakistan to initiate an 'all out war' against militia. However, Pakistan calculated that it will have to pay a huge price. Pakistani military is now in a difficult position to flush out militants since it faces multiple threats not only from Afghanistan but its own territory. Pakistan already faces war or conflict within its own soil. Pakistan has different categories of militia besides widely known TTP, al-Qaeda, Haqqani network but also Sunni to Shia sectarian motivated militia, anti and pro-Indian militant groups such as well sectarian and foreign militants. Sectarian conflict with Balochistan Liberation Army in southwest Pakistan is a battle of different course.

Pakistan faces dozens of major militant groups, which are of deadly and uncivilized nature. There is also a strong criminal gang and internationally organized mafia elements targeting civilians, security personnel to politicians. It is difficult to predict how some of the Pakistani secret elements cut ties with several militant groups. Pakistan military has a high profile threat of retaliation against the militants and could even spread violence in all the major Pakistani cities.

PAKISTAN'S ROLE IN THE WAR ON TERROR

Pakistan's role in the War for Terror was initiated by the September 11 attacks in the United States in 2001 on the World

Trade Center and the Pentagon. These acts were a new manifestation of terrorism, which altogether changed the political psyche of the world. The problem of terrorism, which had been confined to small groups and few states, was changed to a global menace.

The Saudi born Zayn al-Abidn Muhammed Hasayn Abu Zubaydah was arrested by Pakistani officials during a series of joint U.S. and Pakistan raids during the week of March 23, 2002. During the raid the suspect was shot three times while trying to escape capture by military personnel. Zubaydah is said to be a high-ranking al-Qaeda official with the title of operations chief and in charge of running al-Qaeda training camps.

Later that year on September 11, 2002, Ramzi Binalshibh was arrested in Pakistan after a three-hour gunfight with police forces. Binalshibh is known to have shared a room with Mohammad Atta in Hamburg, Germany and to be a financial backer of al-Qaeda operations. It is said Binalshibh was supposed to be another hijacker, however the U.S. Citizenship and Immigration Services rejected his visa application three times, leaving him to the role of financier. The trail of money transferred by Binalshibh from Germany to the United States links both Mohammad Atta and Zacarias Moussaoui.

On March 1, 2003, Khalid Shaikh Mohammed was arrested during CIA-led raids on the suburb of Rawalpindi, nine miles outside of the Pakistani capital of Islamabad. Mohammed at the time of his capture was the third highest ranking official in al-Qaeda and had been directly in charge of the planning for the September 11 attacks.

Escaping capture the week before during a previous raid, the Pakistani government was able to use information gathered from other suspects captured to locate and detain Mohammed. Mohammed was indicted in 1996 by the United States government for links to the Oplan Bojinka, a plot to bomb a series of U.S. civilian airliners.

Other events Mohammed has been linked to include: ordering the killing of Wall Street Journal reporter Daniel Pearl, the USS Cole bombing, Richard Reid's attempt to blow up a civilian airliner with a shoe bomb, and the terrorist attack at the El Ghriba synagogue in Djerba, Tunisia. Khalid Shaikh Mohammed has described himself as the head of the al-Qaeda military committee. Amidst all this, in 2006, Pakistan was accused by NATO commanding officers of aiding and abetting the Taliban in Afghanistan; but NATO later admitted that there was no known evidence against the ISI or Pakistani government of sponsoring terrorism. However in 2007, allegations of ISI secretly making bounty payments up to CDN$ 1,900 (Pakistani rupees. 1 lakh) for each NATO personnel killed surfaced The Afghan government also accuses the ISI of providing help to militants including protection to the recently killed Mullah Dadullah, Taliban's senior military commander, a charge denied by the Pakistani government. India, meanwhile continues to accuse Pakistan's Inter-Services Intelligence of planning several terrorist attacks in Kashmir and elsewhere in the Indian republic, including the July 11, 2006 Mumbai train bombings, which Pakistan attributes it to "homegrown" insurgencies.

Many other countries like Afghanistan and the UK have also accused Pakistan of State-sponsored terrorism and financing terrorism. The upswing in American military activity in Pakistan and neighboring Afghanistan corresponded with a great increase in American military aid to the Pakistan government. In the three years before the attacks of September 11, Pakistan received approximately $9 million in American military aid. In the three years after, the number increased to $4.2 billion, making it the country with the maximum funding post 9/11.

Such a huge inflow of funds has raised concerns that these funds were given without any accountability, as the end uses not being documented, and that large portions were used to suppress civilians' human rights and to purchase weapons to contain domestic problems like the Balochistan unrest.

Waziristan

In 2004 the Pakistani Army launched a campaign in the Federally Administered Tribal Areas of Pakistan's Waziristan region, sending in 80,000 troops. The goal of the conflict was to remove the al-Qaeda and Taliban forces in the region.

After the fall of the Taliban regime many members of the Taliban resistance fled to the Northern border region of Afghanistan and Pakistan where the Pakistani army had previously little control. With the logistics and air support of the United States, the Pakistani Army captured or killed numerous al-Qaeda operatives such as Khalid Shaikh Mohammed, wanted for his involvement in the USS Cole bombing, Oplan Bojinka plot and the killing of Wall Street Journal reporter Daniel Pearl.

However, the Taliban resistance still operates in the Federally Administered Tribal Areas under the control of Haji Omar.

Training Ground for European Militants

In 2009, a politically instable Pakistan emerged as a new global hub for anti-West militancy, but, because of the constant threat of US attacks, recruits were reportedly more likely to spend their time under instruction and in training than carrying out assertive action. In his report on the matter, focusing on an alarming influx of European extremists, *Reuters* security correspondent William Maclean wrote, Long a favored destination of British militants of Pakistani descent, Pakistan's northwestern tribal areas are now attracting Arabs and Europeans of Arab ancestry who three years ago would probably have gone to Iraq to fight U.S. forces.

With the Iraq war apparently winding down, security sources say, the lure for these young men is to fight U.S. forces in neighboring Afghanistan or to gain the skills to carry out attacks back home in the Middle East, Africa or the West. One consequence: Western armies in Afghanistan increasingly face the possibility of having to fight their own compatriots. He added that the matter was likely to surface in a meeting on May 6 between United States President Barack Obama, Pakistani President Asif Ali Zardari and Afghan President Hamid Karzai, the first-mentioned looking to bring an end to the employment of Pakistan's tribal zones as a launching pad for al Qaeda activity around the world.

Background

The events of September 11 impacted international polices and the regional situation within Pakistan necessitated change to its internal policies. Pakistan found that it had no risk-free options: all polices were full of danger and risk of varying degrees. The test was to adopt such a policy and course of action that could minimize the risk and offered the best possible option in the given circumstances.

Geo-strategic location of Pakistan and links with Taliban administration absolved Pakistan to remain unaffected immediately after the attack. Moreover Pakistan was among the three countries, which recognized the Taliban government. Any effort of US and World coalition against Taliban could not have been succeeded without active cooperation of Pakistan. After declaring Al-Qaeda and Osama bin Laden as a prime suspect U.S. President George W. Bush said,

"Every nation in every region now has a decision to make. Either you are with us or with the terrorists."

The Taliban were asked to hand over Osama and close down bases of his Al-Qaeda network or face the consequences. The rest of the world was told that there could not be any neutral party in the war against terrorism. Pakistan, due to its strategic importance and close relation with the Taliban regime, was asked to cooperate. Military President Pervez Musharraf, was told to either abandon the support of Taliban or be prepared to be treated like the Taliban. The military government, due to Pakistan's compulsions or concerns, Pakistan security and stability from external threat, the revival of economy, its nuclear and missile assets and Kashmir cause decided to join US led coalition on war against terrorism. So once again, under another military ruler, Pakistan became front line state.

Actions

After joining the war on terrorism and abandoning its Taliban policy Pakistan opted for a very crucial and challenging task. Though Musharaf tried skilfully to manage the political dissent inside the country, it so far has found it very hard to manage law and order situation in the country. Pakistan has played a vital role in the war against terrorism. It has been a key ally in this war and suffered a lot. Despite its enormous efforts in war against terrorism,

Pakistan has been criticized by the US. It is a fact that US cannot win without Pakistan's help, as 75 present of US/NATO supplies pass through Pakistan and the country has deployed more than 120,000 of its troops in tribal areas.

It has lost more than three thousand soldiers in the war against terrorism. Pakistan has established 1500 checkpoints along the border with Afghanistan, and is the only US ally which has captured or killed more than 700 Al-Qaeda members. Despite all these efforts Pakistan has been blamed for not doing enough.

EFFECTS OF PAKISTAN'S ROLE IN THE WAR

On the International Stage

Pakistan's role in war against terrorism significantly improved her international status. Scholars argue that, US would leave Pakistan after gaining her objectives in war against terrorism. After joining the war on terror Pakistan gained billions of dollars of economic and military aid and also made some gains diplomatically but has lost its control over Afghan affairs.

Within Pakistan

After September 11 and US invasion in Afghanistan there has been a dramatic rise in domestic terrorism, extremism and sectarianism in Pakistan. Pakistan is in the grip of regular suicide bombings, and has suffered high levels of civilian and military casualties as a result.

Baluchistan and Tribal area issues are other implications for Pakistan, with US drones attacks in tribal areas being a real source of concern for Pakistani citizens. The US so far has carried out number of attacks in Pakistani territory. These attacks have led to deterioration of the security situation in tribal areas and the rest of the North-West Frontier Province. In Pakistan there is a growing resentment among people against these attacks on Pakistani soil.

Western Involvement in Terrorism Inside Pakistan

On July 8, 2010, Pasha categorically stated that foreign powers are involved in terrorist activities inside Pakistan. He said, ""The foreign powers are involved in terrorism and destabilization of the country." During a briefing in the National Security Committee session headed by Senator Raza Rabbani, DG ISI Ahmed Shuja Pasha said that the western powers are involved in the terror activities inside the country.

"The US policy against terrorism is under consideration and the changes will be brought with time in accordance with the national interest," he said. This is not new claim inside Pakistan, many Pakistani political commentators have been speaking publicly about this phenomenon.

It is alleged that terrorist activities inside Pakistan were actively carried out by the CIA, the RAW and the MOSSAD for different purposes. Chief among those, is changing the public perception of the War on Terror in the favor of the United States. There are confirmed reports that to achieve its objectives the CIA hired the services of at least a dozen Afghan warlords inside Afghanistan and provided through them arms and finances to militants in FATA and Swat to carry out murders and devastation in the country. It was like a double-edged sword not only to get the Army launch attacks against Taliban on Pakistani side of the border but also to give a message to the ISI that the CIA can use the Pakistani Taliban against their own security forces. Thousands of people have been killed in Pakistan due to terror attacks since the beginning of war on terror.

Future Concerns

There is no doubt that a military campaign is necessary against extremists in the country but it must be supported by social, economic and political change with political reforms being at the top of agenda. The future of Pakistan's role in the war against terrorism is dependent on its political and economic ties with the US. Pakistan is a part of global war on terrorism and was not only defending itself, but also protecting the entire world form the catastrophe of terrorism and extremism, therefore, needs strategic partnership with the rest of the world to foster peace and stability in the region, promote economic stability and address energy needs.

WAR IN NORTH-WEST PAKISTAN

The War in North-West Pakistan is an armed conflict between the Pakistan Armed Forces and armed religious groups such as the Tehrik-i-Taliban Pakistan (TTP), Lashkar-e-Islam, TSNM, Arab and Central Asian militants including Al-Qaeda, regional armed movements and elements of organized crime. The armed conflict began in 2004 when tensions, rooted in the Pakistan Army's search for Al Qaeda fighters in Pakistan's mountainous Waziristan area (in the Federally Administered Tribal Areas), escalated into armed resistance.

Pakistan's actions were presented as Pakistan's role in the War on Terror in the international War on Terrorism. Clashes erupted between Pakistani army troops and Arab and Central Asian militia forces. The foreign militants were joined by Pakistani non-military veterans of the War in Afghanistan (2001–present) which subsequently established the Tehrik-i-Taliban Pakistan and other militia organizations such as Lashkar-e-Islam. The Tehreek-e-Nafaz-e-Shariat-e-Mohammadi (TSNM) established in 1992 allied with the Tehrik-i-Taliban Pakistan and Lashkar-e-Islam.

Background

December 2003: Waziri attempts on President Musharraf's life

In December 2003, two assassination attempts against President Pervez Musharraf were traced to Waziristan. The government responded by intensifying military pressure on the area, however the fighting was costly and government forces would sustain heavy casualties throughout 2004 and into early 2005 when the government switched to a tactic of negotiation instead of direct conflict.

2004: Fighting Breaks Out

On March 16, 2004, the bloody mountainous battle between Pakistan Army and the combined Taliban and Al-Qaeda fighters was insued. Both the International and National media speculated that Pakistan Armed Forces had surrounded the *"High Value Target"* in the mountainous region, possibly Al-Qaeda's Second-in-Command Ayman al-Zawahiri. The battle was concluded as the

Army had captured the entire mountainous region, and captured hundreds of Al-Qaeda fighters. However, the Army was failed to captured Ayman al-Zawahiri as he was either later escaped or he was never among the fighters.

Peace Deals

In April 2004, the Government of Pakistan signed the first of three peace agreements with militants in South Waziristan. It was signed with militia commander Nek Muhammad Wazir, but was immediately abrogated once Nek Muhammad was killed by American Hellfire missile in June 2004. The second was signed in February 2005 with Nek's successor Baitullah Mehsud, which brought relative calm in the South Waziristan region. This deal will be later mimicked in the neighboring North Waziristan territory in September 2006 as the third and final truce between the government and the militants. However, all of these truces would not have a substantial effect in reducing bloodshed. The later two deals were officially broken in August 2007 with the Lal Masjid siege, which raised the suicide attacks on Pakistani forces tenfold throughout the country.

2005-2006

On May 4, 2005, Pakistani commandos captured Abu Faraj al-Libbi after a raid outside the town of Mardan, 30 miles northeast of Peshawar. Abu Farraj al-Libbi was a high ranking al-Qaeda official, rumored to be third after Osama bin Laden and Ayman al-Zawahiri. Al-Libbi replaced Khalid Shaikh Mohammed after his arrest in March 2003 in connection with the September 11th attacks. The Pakistani government arrested al-Libbi and held him on charges in relation to being a chief planner in two assassination attempts on the life of President Pervez Musharraf in December 2003.

On January 13, 2006, the U.S. launched an airstrike on the village of Damadola. The attack occurred in the Bajaur tribal area, about 7 km (4.3 mi) from the Afghan border, and killed at least 18 people. The attack again targeted Ayman al-Zawahiri, but later evidence suggests he was not there.

June 2006: Ceasefire

On June 21, 2006, pro-Tehrik-i-Taliban Pakistan militants in the Bannu region of North Waziristan stated they shot down a Bell military helicopter that was reported to have crashed. The government denied missile fire as the cause, stating it was due to technical faults. The helicopter had taken off from a base camp in Bannu at around 7am for Miramshah and crashed 15 minutes later into the Baran Dam in the Mohmandkhel area on Wednesday morning. Four soldiers were killed including Captain Shams (pilot), Captain Faisal (co-pilot), Lance Naik Ikram and sepoy Altaf, while three other, Sardar Ali, Waseem Abbas and Mohammad Arshad, were rescued. On the same day militants killed an inspector and two constables on a road connecting Bannu and the main town of Miranshah. On June 21, 2006, Afghan Taliban leader Sirajuddin Haqqani issued a decree that it was not (Afghan) Taliban policy to fight the Pakistan Army. However, the Tehrik-i-Taliban Pakistan intentionally did not circulate the decree in North Waziristan thereby keeping pressure on the government.

September 2006: Waziristan Peace Accord Signed

On September 5, 2006, the Waziristan Accord, an agreement between tribal leaders and the Pakistani government was signed in Miranshah, North Waziristan. to end all fighting. The agreement includes the following provisions:

- The Pakistani Army will help reconstruct infrastructure in tribal areas of North and South Waziristan.
- The Pakistani Military will not tolerate any assistance to intruders in North Waziristan, and will monitor actions in the region.
- The Pakistan government is to compensate tribal leaders for the loss of life and property of innocent tribesmen.
- "Foreigners" (informally understood to be foreign jihadists) are not allowed to use Pakistani territory for any terrorist activity anywhere in the world.
- 2,500 foreigners who were originally held on suspicion of having links to the Taliban were to be detained for necessary action against them.

The agreement, dubbed the Waziristan accord, has been viewed by some political commentators as a success for Pakistan. Further details of the agreement, as well as comments on the agreement made by US, Pakistani, and Taliban spokesmen is available in the Waziristan accord article.

October 2006: The Madrassa Air Strike

On October 30, 2006, the Pakistani army conducted an air strike against a madrassa in the Bajaur region bordering Afghanistan. The madrassa was destroyed killing 70 to 80 people. In retaliation for the attack the militants conducted a suicide bombing on an army camp on November 8, 2006, killing 42 Pakistani soldiers and wounding 20.

2007

In March, Pakistan signed a peace treaty with Faqir Mohammed, the main militant leader in Bajaur. Militant groups now held three districts in the Federally Administered Tribal Areas: South Waziristan, North Waziristan and Bajaur Agency.

Waziri-Uzbek Tensions

Reportedly, the fighting sparked by the killing of Saiful Adil, an al-Qaeda-linked Arab, blamed on the Uzbeks by Maulavi Nazir, described as a top pro-Taliban militant commander in the region. According to the other version, fighting started after Mullah Nazir, whom the government says has come over to its side, ordered the Uzbek followers of Tohir Yo'ldosh, formerly a close confidant of Osama bin Laden, to disarm. It was also preceded by the clashes between the IMU and a pro-government tribal leader in Azam Warsak, in which 17 to 19 people died before a ceasefire was announced.

Defeat of the Islamic Movement of Uzbekistan

Local militants allied to the tribesmen were reported attacking and seizing the IMU's private jail in Azam Warsak. The Pakistan Army said it did not intend to step in, but witnesses say government artillery fired on the Uzbek bunkers they set up to fight the tribesmen. Heavy fighting resumed on March 29, ending a week-long ceasefire between tribal fighters and foreign militants. According to initial reports, tribesmen attacked a checkpoint manned by Uzbek militants and captured two of them. The clashes also left one tribal fighter dead and three wounded.

The following day, a senior Pakistani official announced that 52 people were killed during the past two days, 45 of them Uzbeks and the rest tribesmen. One of Maulvi Nazir's aides put the death toll at 35 Uzbeks and 10 tribal fighters.

However, residents in the area said that the death toll on both sides was inflated. The conflict further escalated on April 2 when a council of elders declared jihad against foreign militants and started to raise an army of tribesmen. According to Pakistani security officials, heavy fighting concentrated in the village of Doza Ghundai left more than 60 people dead, including 50 foreigners, 10 tribal fighters and one Pakistani soldier. He also said that "dozens" of Uzbeks had surrendered to tribal forces and that many bunkers used by militants were seized or destroyed. On April 12, 2007 the army general in charge of South Waziristan said that tribal fighters had cleared the Uzbeks out of the valleys surrounding Wana and the foreign militants had been pushed back into the mountains on the Afghan border. Four days later, the local tribesmen has urged Islamabad to resume control of law and order in the area.

Lal Masjid Siege

On July 3, 2007, the militant supporters of Lal Masjid and Pakistani security forces clashed in Islamabad after the students from the mosque attacked a nearby government ministry building with stones. Their resultant faceoff with the military escalated, despite the intervention of then-ruling Pakistan Muslim League (Q) leaders Chaudhry Shujaat Hussain and Muhammad Ijaz-ul-Haq. The Pakistani security forces immediately put up a siege around the mosque complex which lasted until July 11 and resulted in 108 deaths. This represented the main catalyst for the conflict and eventual breakdown of the truce that existed between Pakistan and the Pakistani Taliban groups. Already during the siege there were several attacks in Waziristan in retaliation for the siege.

Truce in Waziristan Broken

As the siege in Islamabad ensued, several attacks on Pakistani troops in Waziristan were reported. On July 14, 2007, a suicide bomber attacked a Pakistani Army convoy killing 25 soldiers and wounding 54. On July 15, 2007, two suicide bombers attacked another Pakistani Army convoy killing 16 soldiers and 5 civilians and wounding another 47 people. And in a separate incident a fourth suicide bomber attacked a police headquarters killing 28 police officers and recruits and wounding 35 people. The assault on the Red Mosque prompted rebels along the border with Afghanistan to scrap the controversial Waziristan Accord with the government.

The new war in Waziristan

The Army moved large concentration of troops into Waziristan and engaged in fierce clashes with militants in which at least 100 militants were killed including wanted terrorist and former Guantanamo Bay detainee, Abdullah Mehsud. The militants also struck back by attacking Army convoys, security check points and sending suicide bombers killing dozens of soldiers and police and over 100 civilians. In one month of fighting during the period from July 24 to August 24, 2007, 250 militants and 60 soldiers were killed. On September 2, 2007, just a few dozen militants led by Baitullah Mehsud managed to ambush a 17-vehicle army convoy and captured an estimated 247 soldiers without a shot being fired, an event that shocked the nation. Several officers were among the captured.

After the army returned to Waziristan, they garrisoned the areas and set up check-points, but the militants hit hard. In mid-September the Tehrik-i-Taliban Pakistan and other forces attacked a number of Pakistani army outposts all across North and South Waziristan. This resulted in some of the heaviest fighting of the war. Following the Lal Masjid Siege, On September 12, 2007, the first outpost was attacked and overrun by the militants resulting in the capture of 12 Pakistani soldiers.

The next day on September 13, 2007, a suicide bomber in Tarbela Ghazi attacked a Pakistani army base, destroying the main mess hall and killing 20 members of the Karrar commando group; Pakistan's most elite army unit. A series of attacks ensued and by September 20, 2007 a total of five Pakistani Army military outposts had been overrun and more than 25 soldiers captured. More than 65 soldiers were either killed or captured and almost 100 wounded. A little over two weeks later, the Army responded with helicopter gunships, jet fighters and ground troops. They hit militant positions near the town of Mir Ali. In heavy fighting over four days, 257 people were killed, including 175 militants, 47 soldiers and 35 civilians.

Battle of Swat Valley

By the end of October fighting erupted in the Swat district of the Khyber Pakhtunkhwa province, with a large Tehreek-e-Nafaz-e-Shariat-e-Mohammadi force, under the command of Maulana Fazlullah, trying to impose Sharia law. Around 3,000 paramilitary soldiers were sent to confront them. After almost a week of heavy fighting the battle came to a standstill with both sides suffering heavy casualties. Then on November 1 and November 3, 220 paramilitary soldiers and policemen surrendered or deserted after a military position on a hill-top and two police stations were overrun.

This left the Tehreek-e-Nafaz-e-Shariat-e-Mohammadi (TNSM) in control of most of the Swat district. The fighting in Swat is the first serious insurgent threat from militia forces in what is known as a settled area of Pakistan. Forces loyal to Maulana Fazlullah, including some foreign fighters, after taking control of a series of small towns and villages, tried to implement strict Islamic law in November 2007. In mid-November the regular army was deployed with the help of helicopter gunships to crush the uprising. The Pakistan Army deployed over 2,500 men. By the beginning of December the fighting had ended and the Army recaptured Swat. Almost 400 of Maulana Fazlullah's fighters were dead along with 15 Pakistani soldiers and 20 civilians in the military offensive. Despite the victory by the Pakistani army, TNSM militants slowly re-entered Swat over the coming months and started engaging security forces in battles that lasted throughout 2008. By early February 2009, the whole district was in Pakistan Army Control.

The Rawalpindi Attacks

On September 3, two suicide bombers targeted a military intelligence (ISI) bus and a line of cars carrying ISI officers. The bus attack killed a large number of Defence Ministry workers and the other attack killed an Army colonel. In all 31 people, 19 soldiers and 12 civilians, were killed. Over two months later on November 24, one of the targets was a military intelligence bus. Almost everyone on the bus was killed. The other bomber blew up at a military checkpoint. 35 people were killed, almost all military.

State of Emergency

The 2007 Pakistani state of emergency was declared by Pervez Musharraf on 2007-11-03 and lasted until 2007-12-15. During this time the constitution of the country was suspended. This action and its responses are generally related to the controversies surrounding the re-election of Musharraf during the presidential election that had occurred on 2007-10-06, and also was claimed by the government to be the reaction to the actions by militants in Waziristan.

Benazir Bhutto's Assassination

On December 27, 2007, Pakistani opposition leader Benazir Bhutto was killed upon leaving a political rally for the Pakistan Peoples Party (PPP) in Rawalpindi, Pakistan. A suicidal assassin reportedly fired shots in Bhutto's direction just prior to detonating an explosive pellet-laden vest, killing approximately 24 people and wounding many more. Musharraf and the army blamed the attack on Al-Qaida, but the following day a statement by Baitullah Mehsud was sent to the media saying that he and Al-Qaida had no involvement in the murder of the former Prime Minister, he briefed that these were the crimes of Musharraf and the army.

The killing was followed by a wave of violence across the country that left 58 people dead, including four police officers. Most of the violence was directed at Musharraf and the pro-Musharraf political party, Pakistan Muslim League (Q). The public chanted slogans against the army and Musharraf: "Musharraf Dog", "General is a murderer", "uniform (army) wearing murderers", etc. Bhutto had previously survived an assassination attempt made on her life during her homecoming which left 139 people dead and hundreds wounded.

2008: More fighting in South Waziristan

In January 2008, militants overran Sararogha Fort, and may have overrun a fort in Ladah as well. Both forts are in South Waziristan, and were held by the Pakistani army. On February 25, 2008, a suicide bomber struck in the garrison-town of Rawalpindi killing Pakistani Lt. Gen. Mushtaq Baig along with two more soldiers and five civilians. Baig was the highest-level military official to be assassinated since Pakistan joined the war on terror.

Operation Zalzala

A full-fledged security operation called 'Zalzala' (*earthquake*) by Pakistan Army's 14th Infantry Division in January to flush out Baitullah Mehsud's Tehrik-i-Taliban Pakistan militants from the area. Until then the area was infested with militants, with some villagers providing them support and shelter. Many militants were killed during the operation, and within three days the security forces were in full control of the area. The army later captured a few other villages and small towns to put the squeeze on Baitullah Mehsud. However, the operation led to a huge displacement of local population. According to 14 Division GOC Maj Gen Tariq Khan, about 200,000 people, including men, women and children, were displaced. Khalid Aziz, former NWFP chief secretary and expert on tribal affairs, said the displacement was "one of the biggest in tribal history" adding that human cost of the conflict in Waziristan "has gone unrecorded."

Peace agreement

On February 7, 2008, the Tehrik-i-Taliban Pakistan (TTP) led by Baitullah Mehsud offered a truce and peace negotiations resulting in a suspension of violence. On May 21, 2008 Pakistan signed a peace agreement with the Tehrik-i-Taliban Pakistan (TTP). Despite the agreement sporadic fighting continued until late June and escalated with the takeover of the town of Jandola on June 24, by the militants. 22 pro-government tribal fighters were captured and executed by the TTP at that time. On June 28, 2008, Pakistan's Army started an offensive against militia fighters in Khyber, codenamed *Operation Sirat-e-Mustaqeem*. The military took control of a key town and demolished an insurgent group's building. 1 militant was reportedly killed while 2 soldiers died in Swat valley. The operation was halted in early July.

On July 19, 2008, clashes erupted between the Tehrik-i-Taliban Pakistan and a faction of pro-government Taliban militants. 10-15 of the pro-government fighters were killed and another 120 were captured. Among the captured were two commanders who were tried under "Islamic" law by the Taliban and then executed. On July 21, 2008, heavy fighting in Baluchistan killed 32 militants, 9 soldiers and 2 civilians. More than two dozen militants were captured and a large weapons cache was found. Between July 28 and August 4, 2008, heavy fighting flared up in the northwestern Swat valley leaving 94 militants and 22 soldiers and policemen dead. Another 28 civilians were also killed.

Bajaur Offensive Starts

Heavy fighting erupted on August 6, 2008, in the Loisam area of Bajaur district. Loisam lies on the strategically important road leading towards the main northwestern city of Peshawar. The fighting started when hundreds of militants poured into the area attacking government forces. After four days of fighting on August 10 the military withdrew from the area. 100 militants and 9 soldiers were confirmed killed and another 55 soldiers were missing, at least three dozen of them captured by the militants. While the fighting was going on in Bajaur, in the Buner area of Khyber-Pakhtunkhwa militants killed at least nine policemen in an attack on a check post.

The checkpoint was then abandoned, and the local Pakistani forces withdrew to Khar, the main town of Bajaur Agency. There were reports that the town of Khar was then besieged by tribal militants. On August 21, 2008, in response to the military offensive in Bajaur, two suicide bombers attacked the Pakistan Ordnance Factories in Wah while workers were changing shifts. The attack killed at least 70 people.

Tribesmen Declare War Against the Militants

By the beginning of September 2008, Pakistani tribal elders began organising a private army of approximately 30,000 tribesmen to fight the militias. A lashkar, or private army, composed of Pakistani tribesmen, began torching the houses of militia commanders in Bajaur, near the Afghan border, vowing to fight them until they are expelled. This included the house of a local militant commander named Naimatullah, who had occupied several government schools and converted them into seminaries. A tribal elder named Malik Munsib Khan, who heads the lashkar, said that tribesmen would continue their struggle until the militants were expelled from the area, adding that anyone found sheltering militants would be fined one million rupees and their houses will be torched. The tribesmen also torched two important centres of the militants in the area and gained control of most of the tehsil.

The main reasons for this was that the operations that were taking place in the Federally Administered Tribal Areas had displaced some 300,000 people while dozens of citizens have been killed in clashes between the militants and military. Since the start of Pakistan's war against the militants some 150,000 tribesmen have sided with them.

US Support and Aid for Pakistani Tribesmen

Recent American military proposals outlines an intensified effort to enlist tribal leaders in the frontier areas of Pakistan in the fight against Al Qaeda, the Afghan Taliban and Pakistani militia groups, as part of a broader effort to bolster Pakistani forces against militancy in the region. The proposal is modeled in part on a similar effort by American forces in Iraq that has been hailed as a great success in fighting foreign insurgents there. But it raises the question of whether such partnerships can be forged without a significant American military presence in Pakistan. And it is unclear whether enough support can be found among the tribes. Small numbers of United States military personnel have served as advisers to the Pakistani Army in the tribal areas, giving planning advice and helping to integrate American intelligence. Under this new approach, the number of advisers would increase. American officials said these security improvements complemented a package of assistance from the Agency for International Development and the State Department for the seven districts of the tribal areas that amounted to $750 million over five years, and would involve work in education, health and other sectors. The State Department's Bureau of International Narcotics and Law Enforcement Affairs is also assisting the Frontier Corps with financing for counternarcotics work.

Islamabad Marriott Hotel Bombing

On September 23, 2008, the Pakistani Army, backed by helicopter gunships and artillery killed more than 60 insurgents in northwest Pakistan in offensives as the response to the Islamabad Marriott Hotel bombing over the weekend at the Marriott hotel in the capital Islamabad that killed 53 people. In the nearby Bajur tribal region, the Army killed at least 10 militants, according to government officials. The Bajur operations, which the army says has left more than 700 suspected militants dead, has won praise from U.S. officials.

Renewed Bajaur Offensive

Pakistani President Asif Ali Zardari publicly vowed revenge in response to the Islamabad Marriott Hotel bombing. By September 26, 2008, Pakistani troops had successfully conducted and completed a major offensive in the Bajaur and the Tang Khata regions of the Federally Administered Tribal Areas, codenamed *Operation Sherdil*. Pakistani troops had killed over 1,000 militants in a huge offensive, a day after President Asif Ali Zardari lashed out at US forces over a clash on the Afghan border.

Tariq Khan, Inspector General of the Paramilitary Frontier Corps, mentioned to journalists that since the beginning of the Bajaur operations, there were up to 2,000 militant fighters including hundreds of foreigners who were fighting with the soldiers and the security forces. The overall death toll was over 1,000 militants and also adding that 27 Pakistani soldiers had also been killed with 111 soldiers seriously wounded. Five top Al-Qaeda and militia commanders were among those killed in a month-long operation in Bajaur. Of the five militant commanders killed, four appeared to be foreigners: Egyptian Abu Saeed Al-Masri; Abu Suleiman, an Arab; an Uzbek commander named Mullah Mansoor; and an Afghan commander called Manaras. The fifth was a Pakistani commander named only Abdullah, a son of ageing hardline leader Maulvi Faqir Mohammad who is based in Bajaur and has close ties to Al-Qaeda second-in-command Ayman al-Zawahiri.

Between October 22 and October 24, security forces engaged in another push against militants in the restive Bajaur and Khyber tribal regions. Air strikes were carried out in the Nawagai and Mamond sub-districts of Bajaur Agency. The troops destroyed several centres of militants at Charmang, Chinar and Zorbandar and had inflicted heavy losses on them. Gunship helicopters shelled in Charming, Cheenar, Kohiand Babarha areas of Nawagai and Mamund Tehsil of Bajaur agency, destroying various underground hideouts and bunkers of militants. The security forces had also taken control of different areas of Loisam, a militant headquarters, and advanced towards other areas for complete control.

Intensified U.S. Strikes

Since the end of August 2008, the United States had stepped up its attacks in the Federally Administered Tribal Areas. On September 3 a commando attack took place in a village near the Afghan border in South Waziristan, and there have been 3 strikes from unmanned drones in North Waziristan, culminating on the morning of September 8, 2008, when a United States Air Force drone aircraft fired a number of missiles at a "guest house for militants arriving in North Waziristan", which unsuccessfully targeted Jalaluddin Haqqani, killed 23 people.

On September 25, 2008, following exchanges of gunfire between US and Pakistani forces on the frontier, President Zardari told the United Nations that Pakistan would not tolerate violations of its sovereignty, even by its allies. The incident happened after two US military helicopters came under fire from the Pakistani side, a US military spokesman said, insisting that they had been about a mile and a half inside Afghanistan.

President Zardari told the United Nations, "Just as we will not let Pakistani's territory to be used by terrorists for attacks against our people and our neighbours, we cannot allow our territory and our sovereignty to be violated by our friends," he said, without citing the United States or the border flareup.

Militants Targeting of Tribesmen

On October 10, 2008, Islamic militants beheaded four kidnapped pro-government tribal elders in the Charmang area of Bajaur on Friday. The militants had killed them because the elders had been pro-government.

On October 11, 2008, a suicide bomber struck an anti-Militant gathering of tribal elders, just as they had decided to form a lashkar (tribal militia), killing at least 110 anti-Taliban tribesmen and wounding 125 others. The suicide bomber drove his car into the gathering itself and blew himself up. The attack on the tribal council took place in Orakzai, normally a relatively quiet corner of the nation's restive tribal areas.

Fighting for the NATO Supply Lines

On October 19, 2008, the Pakistan Army was locked in a fierce battle with militants to keep open the fuel and arms supply routes to British and American forces in Afghanistan. For months, militants had been trying to either attack or seal off the supply routes. The army claimed that Mohammad Tariq Alfridi, the militant commander, had seized terrain around the mile-long Kohat tunnel, south of Peshawar, three times since January. He had coordinated suicide bomb attacks and rocket strikes against convoys emerging from it. Maulvi Omar, a Tehrik-i-Taliban Pakistan spokesman, said that his fighters would lay down their arms if the Pakistan Army ceased fighting. The Pakistan Army ignored his offer. The battle for the tunnel began at the start of the year when Tehrik-i-Taliban fighters seized five trucks carrying weapons and ammunition. They held the tunnel for a week before they were driven out in fierce fighting with the Army. Since then, Tariq and his men have returned several times to attack convoys. The army launched its latest onslaught after a suicide bomb attack at one of its bases near the tunnel six weeks ago. Five people were killed and 45 were injured, including 35 soldiers, when a pickup truck packed with explosives was driven into a checkpoint.

On November 11, 2008, militants attacked two convoys at the Khyber Pass capturing 13 trucks which were headed for Afghanistan. One convoy was from the United Nations World Food Programme and was carrying wheat. The second was intended for NATO troops and one of the captured trucks was carrying with it two U.S. military Humvees, which were also seized. On December 8, 2008, militants torched more than 160 vehicles destined for US-led troops in Afghanistan. The militants attacked the Portward Logistic Terminal in the northern city of Peshawar at around 02:30 AM, destroying its gate with a rocket-propelled grenade and shooting dead a guard. They then set fire to about 100 vehicles, including 70 Humvees, which shipping documents showed were being shipped to the US-led coalition forces and the Afghan National Army. At the same time, militants torched about 60 more vehicles at the nearby Faisal depot, which like Portward is on the ring road around Peshawar, where convoys typically stop before heading for the Khyber Pass.

2009

On February 3, 2009, militants blew up a bridge at the Khyber Pass, temporarily cutting a major supply line for Western troops in Afghanistan. After the attack supplies along the route had been halted "for the time being", according to NATO.

Swat Ceasefire

Pakistan agreed to impose Sharia law and suspend military operations in the scenic Swat Valley. The decision is troubling for the United States, which believes that it will embolden militants who are fighting US-led troops in Afghanistan and want to impose Islamic law across nuclear-armed Pakistan. US officials believe that it will now provide another safe haven for the militants within 80 miles of Islamabad, the Pakistani capital, as well as a corridor between the Afghan border and the disputed region of Kashmir. Pakistani officials said that it was the only way to pacify a fierce Islamist insurgency and avoid more civilian casualties in Swat – whose ski resort and mountain scenery once made it a popular tourist destination. Amir Haider Khan Hoti, the chief minister for the Khyber Pakhtunkhwa province, announced the government's decision after a meeting with militant leaders in the provincial capital, Peshawar. He said that local authorities would impose Islamic law across Malakand region, which includes Swat. Officials said that Asif Ali Zardari, the Pakistani President, would sign off on the deal once peace had been restored. The agreement came the day after the militants in Swat said that it would observe a ten-day ceasefire in support of the peace process. Pakistani officials say that the laws allow Muslim clerics to advise judges, but do not outlaw female education, music or other activities once banned by the Afghan Taliban in Afghanistan.

Defeat of the Militants in Bajaur

On March 1, 2009, the Pakistan Army finally defeated Bajaur militants and foreign militants in Bajaur, which is a strategically important region on the Afghan border. Major-General Tariq Khan, who was commanding the military operations in five of the seven agencies, said his Army and the Frontier Corps had killed most militants in Bajaur, the smallest of the agencies but a major infiltration route into Afghanistan, after a six-month offensive. By the time the six-month long battle in Bajaur was over, the Pakistan Army killed over 1,500 militants while losing 97 of their own soldiers and another 404 soldiers seriously injured.

Militant Counter-attack

On March 30, militia commandos struck in Lahore. They attacked the Munawan Police Academy killing and taking hostage police cadets. A siege was under way for about eight hours after the militants had barricaded themselves in the academy. Eventually police forces managed to retake the compound. 18 people were killed in the attack: eight policemen, eight militants and two civilians. At least 95 policemen were wounded and 10 were taken hostage before being rescued. Four gunmen were captured by the police. On April 4, a suicide bomber attacked a military camp in Islamabad killing eight soldiers. Less than 24 hours later on April 5, two more suicide attacks occurred. One bomber targeted a market on the border with Afghanistan killing 17 people and the other attacked a mosque in Chakwal, in the eastern Pakistani province of Punjab, killing 26 more civilians. The next day, the leader of the Tehrik-i-Taliban Pakistan, Baitullah Mehsud, promised that there were to be two suicide attacks per week in the country until the Pakistani army withdraws from the border region and the United States stops its missile attacks by unmaned drones on militant bases.

Militants Violate Swat Deal

In March 2009, many Pakistanis were horrified when a videotape surfaced that showed miliant enforcers publicly whipping a 17-year-old girl in Swat accused of having an affair. The girl had not committed fornication or adultery but was flogged simply because she refused her brother's demand to marry someone of his choosing. Protests broke out all over Pakistan to demonstrate against the flogging. Raja Zafar ul-Haq, a well-respected Pakistani Islamic scholar and political activist said this summary punishment of flogging simply for refusing a marriage proposal was totally un-Islamic and had nothing to do with Sharia. He went on to say that Prophet Muhammad had strictly forbidden the practice of forced marriages and in this case, the girl had not done anything wrong by refusing a marriage proposal.

A five-member team appointed by the Supreme Court investigated the video's origins, and concluded that it had been faked, raising questions at Pakistani intelligence agencies. In Buner, the Taliban continued their criminal activities when residents said Taliban fighters had been stealing cattle for meat, stealing other livestock, berating men without beards and recruiting teenagers into their ranks. The Taliban also began to steal vehicles belonging to government officials and ransacked the offices of some local non-government organisations for no apparent reason. 12 schoolchildren were killed by a bomb contained in a football.

Second Battle of Swat

On April 26, 2009, the Pakistani Army started Operation Black Thunderstorm, with the aim of retaking Buner, Lower Dir, Swat and Shangla districts from the Pakistani Taliban after the militants took control of the area since the start of the year. The operation largely cleared the Lower Dir district of militia forces by April 28 and Buner by May 5. On May 5, operations started to retake Swat and later on Shangla. Fighting in Swat was particularly fierce since the Taliban threw away their insurgent tactics and the army their counter-insurgency tactics. Both sides favoured more conventional frontline warfare as a means of fighting each other. By May 14, the military was only six kilometers south of Mingora, the milita-held capital city of Swat, and preparations for all-out street fighting was under way.

On May 23, the battle for Mingora started and by May 27, 70 percent of the city was cleared of militants. On May 30, the Pakistani military had taken back the city of Mingora from the Pakistani Taliban, calling it a significant victory in its offensive against the militants. However, some sporadic fighting was still continuing on the city's outskirts. In all, according to the military, 128 soldiers and more than 1,475 militants were killed and 317 soldiers were wounded during operation Black Thunderstorm. 95 soldiers and policemen were captured by the militants, 18 of them were rescued while the fate of the others remained undetermined. 114 militants were captured, including some local commanders. At least 23 of the militants killed were foreigners.

Sporadic fighting throughout Swat continued up until mid-June. On June 14, the operation was declared over and the military had regained control of the region. Only small pockets of Taliban resistance remained and the military started mopping up operations.

This led to a refugee crisis, and by August 22, 1.6 million of 2.3 million have returned home according to UN estimates.

Blockade of South Waziristan

On June 16, 2009, in the aftermath of the successful victory and recapture of the entire Swat valley, the Pakistan Army began a massive troop build-up along the southern and eastern borders of South Waziristan. Pakistan was now taking the fight to Mehsud's mountainous stronghold, ordering an expansion of its current offensive against Tehrik-i-Taliban Pakistan fighters in the Swat valley. On Sunday night, denouncing Mehsud as "the root cause of all evils," Owais Ghani, the governor of the Khyber Pakhtunkhwa, said the government has called on the army to launch a "full-fledged" military operation to eliminate Mehsud and his estimated 20,000 men. The crucial battle may prove to be the most difficult that Pakistan's military has faced on its soil in recent years.

Islamabad's decision to launch the offensive against Mehsud signals a deepening of Pakistani resolve against the militants. The army has targeted the Tehrik-i-Taliban Pakistan leader on three separate occasions — in 2004, 2005 and 2008 — but walked away each time after signing ruinous "peace deals" that have only served to embolden Mehsud. But the military appears more determined this time. It also enjoys the backing of a government that has gained public support as the recent wave of terrorist attacks has heightened revulsion against the Taliban.

Killing of Baitullah Mehsud

The leader of Tehrik-i-Taliban Pakistan, Baitullah Mehsud, was killed in a drone attack in early August 2009. This was later confirmed by captured chief spokesman Maulvi Umar. He was replaced by Hakimullah Mehsud.

Militant October–November 2009 Counter-offensive

In early October 2009, the Tehrik-i-Taliban Pakistan started a string of attacks in cities across Pakistan. The goal of the attacks was to show that the militants were still a united fighting force following the death of their leader and to disrupt a planned military offensive into South Waziristan. Places targeted include the U.N. World Food Program offices in Islamabad, Khyber bazaar in Peshawar, Army General Headquarters in Rawalpindi, a market in Shangla, security establishments in Lahore, police stations in Kohat and Peshawar, the International Islamic University in Islamabad, and Pakistan Air Force Complex in Kamra.

The month ended with a car bombing of Meena Bazaar, Peshawar killing 118 civilians. The army then began a ground offensive in South Waziristan. November saw suicide bombings of the Army's National Bank of Pakistan in Rawalpindi, a market in Charsadda and six bombings of Peshawar including the regional headquarters of the ISI and the Judicial Complex.

South Waziristan Offensive

On October 17, the Pakistani Army launched a large-scale offensive in South Waziristan involving 28,000 troops advancing across South Waziristan from three directions.

On October 19, the first town to fall to the Army was Kotkai, the birthplace of the Pakistani Taliban leader Hakimullah Mehsud. However, the next day, the Tehrik-i-Taliban Pakistan re-took the town from the military.

Troops had thrust into Kotkai only to be hit by a determined counteroffensive that killed seven soldiers, including an army major, and wounded seven more. Still, the Army managed to take the town once again on October 24, after days of bombardments.

On October 29, the town of Kaniguram, which was under the control of Uzbek fighters from the Islamic Movement of Uzbekistan, was surrounded. And on November 2, Kaniguram was taken.

On November 1, the towns of Sararogha and Makin were surrounded, and fighting for Sararogha started on November 3. The fighting there lasted until November 17, when the town finally fell to the military. The same day, the town of Laddah was also captured by the Army and street fighting commenced in Makin. Both Sararogha and Laddah were devastated in the fighting. By November 21, more than 570 militants and 76 soldiers were killed in the offensive.

On December 12, 2009, the Pakistan army declared victory in South Waziristan.

2010

Orakzai and Kurram Offensive

In an offensive in Bajaur by Frontier Corps, a militants' stronghold village Damadola was captured and cleared by February 6, 2010. Bajaur was declared conflict free zone by April 20. On March 23, 2010, the Pakistan army launched an offensive to clear Orakzai. Officials also announced a future offensive in North Waziristan. The week prior the Pakistan military killed approximately 150 militants in fighting in the region. It is expected that all tribal areas would be cleared by June 2010. June 3, Pakistani authorities announced a victory over the insurgents in Orakzai and Kurram.

Casualties

2,637 security forces members and 12,847 militants were killed between January 2003, and November 12, 2009, according to government sources.

On February 18, 2010, the Pakistani military confirmed that since September 11, 2001, 2,351 soldiers have been killed in the war and another 6,512 have been wounded. Also, there have been 21,672 civilian casualties (at least 7,598 of these were killed). The military stated that 17,742 militants have been killed or captured. Among these, by November 2007, were 488 foreign extremists killed, 24 others arrested and 324 injured. An additional 220 policemen were killed in fighting in 2007 and 2008. Before all-out fighting broke out in 2003, independent news sources reported only four incidents of deaths of Pakistani security forces members in 2001 and 2002, in which a total of 20 soldiers and policemen were killed. The independent South Asia Terrorism Portal website has estimated that at least 1,865 soldiers and policemen were killed between 2003 and 2008. The Pak Institute For Peace Studies has estimated that 1,185 soldiers and policemen were killed in 2009.

Also, at least 857 soldiers and policemen have been reported captured by the militants since the start of the war, with at least 558 of them being released.

Mr. Naushad Ali Khan Superintendent of Research and Analysis, NWFP Police in his article *Suicide and terrorist attacks and police actions in NWFP, Pakistan* has provided details of different activities of the terrorists during 2008. Accordingly 483 cases were registered with 533 deaths and 1290 injured. Similarly 29 suicidal cases were registered that resulted in the death of 247 persons while 695 persons sustained injuries. During the same period 83 attempts of terrorism were foiled by the NWFP Police. The full article can be viewed on the official website of Pakistan Society of Criminology.

United States Role

Until July 2009 the conflict, as well as terrorism in Pakistan, had cost Pakistan $35 billion. According to US Congress and the Pakistani media, Pakistan has received about $18 billion from the United States for the logistical support it provided for the counter-terrorism operations from 2001 to 2010, and for its own military operation mainly in Waziristan and other tribal areas along the Durand line.

The Bush administration also offered an additional $3 billion five-year aid package to Pakistan for becoming a frontline ally in its 'war on terror'. Annual instalments of $600 million each split evenly between military and economic aid, began in 2005.

In 2009, President Barack Obama pledged to continue supporting Pakistan and has said Pakistan would be provided economic aid of $1.5 billion dollars each year for the next five years. Unfolding a new US strategy to defeat Taliban and Al-Qaeda, Obama said Pakistan must be a 'stronger partner' in destroying Al-Qaeda safe havens. In addition, President Obama has also planned to propose an extra $2.8 billion dollars in aid for the Pakistani military to intensify the US-led war on terror along the Pakistan-Afghanistan border. The military aid would be in addition to the civilian aid of $1.5 billion dollars a year for the next five years from 2009 onwards.

In his autobiography, President Musharraf wrote that the United States had paid millions of dollars to the Pakistan government as bounty money for capturing al-Qaeda operators from tribal areas bordering Afghanistan. About 359 of them were handed over to the US for prosecution. Pakistan has bought 1,000 laser-guided bomb kits and 18 F-16 fighter jets from USA, to attack militants.

Amongst Pakistanis opinion about the role of the US is generally negative. Incidents of terrorism cause rage and anger against the terrorist organizations but they also cause frustration with the United States. According to Pew Global Polls only 17% of Pakistanis have a positive view of the US and only 11% it as a useful partner in the war on terror.

TERRORIST GROUPS, AL-QAEDA AND U.S. POLICY IMPLICATIONS

Al-Qaeda's Fluid organization, methods of recruitment, funding, and means of communication distinguish it as an advancement in twenty-first-century terrorist groups. Al-Qaeda is a product of the times. Yet it also echoes historical predecessors, expanding on such factors as the international links and ideological drive of nineteenth-century anarchists, the open-ended aims of Aum Shinrikyo, the brilliance in public communications of the early PLO, and the taste for mass casualty attacks of twentieth-century Sikh separatists or Hezbollah.

Al-Qaeda is an amalgam of old and new, reflecting twenty-first century advances in means or matters of degree rather than true originality; still, most analysts miss the connections with its predecessors and are blinded by its solipsistic rhetoric. That is a mistake. The pressing challenge is to determine which lessons from the decline of earlier terrorist groups are relevant to al-Qaeda and which are not.

First, past experience with terrorism indicates that al-Qaeda will not end if Osama bin Laden is killed. There are many other reasons to pursue him, including bringing him to justice, removing his leadership and expertise, and increasing esprit de corps on the Western side (whose credibility is sapped because of bin Laden's enduring elusiveness). The argument that his demise will end al-Qaeda is tinged with emotion, not dispassionate analysis. Organizations that have been crippled by the killing of their leader have been hierarchically structured, reflecting to some degree a cult of personality, and have lacked a viable successor. Al-Qaeda meets neither of these criteria: it has a mutable structure with a strong, even increasing, emphasis on individual cells and local initiative. It is not the first organization to operate in this way; and the demise of similar terrorist groups required much more than the death of one person. Unlike the PKK and Shining Path, al-Qaeda is not driven by a cult of personality; despite his astonishing popularity, bin Laden has deliberately avoided allowing the movement to revolve around his own persona, preferring instead to keep his personal habits private, to talk of the insignificance of his own fate, and to project the image of a humble man eager to die for his beliefs. As for a viable replacement, bin Laden has often spoken openly of a succession plan, and that plan has to a large degree already taken effect. Furthermore, his capture or killing would produce its own countervailing negative consequences, including (most likely) the creation of a powerful martyr. On balance, the removal of bin Laden would have important potential benefits, but to believe that it would kill al-Qaeda is to be ahistorical and naive.

Second, although there was a time when the failure to transition to a new generation might have been a viable finale for al-Qaeda, that time is long past. Al-Qaeda has transitioned to a second, third, and arguably fourth generation. The reason relates especially to the second distinctive element of al-Qaeda: its method of recruitment or, more accurately, its attraction of radicalized followers (both individuals and groups), many of whom in turn are connected to existing local networks. Al-Qaeda's spread has been compared to a virus or a bacterium, dispersing its contagion to disparate sites. Although this is a seductive analogy, it is also misleading: the perpetuation of al-Qaeda is a sentient process involving well-considered marketing strategies and deliberate tactical decisions, not a mindless "disease" process; thinking of it as a "disease" shores up the unfortunate American tendency to avoid analyzing the mentality of the enemy.

Al-Qaeda is operating with a long-term strategy and is certainly not following the left-wing groups of the 1970s in their failure to articulate a coherent ideological vision or the peripatetic right-wing groups of the twentieth century. It has transitioned beyond its original structure and now represents a multigenerational threat with staying power comparable to the enthonationalist groups of the twentieth century. Likewise, arguments about whether al-Qaeda is best described primarily as an ideology or by its opposition to foreign occupation of Muslim lands are specious: al-Qaeda's adherents use both rationales to spread their links. The movement is opportunistic. The challenge for the United States and its allies is to move beyond rigid mind-sets and paradigms, do more in-depth analysis, and be more nimble and strategic in response to al-Qaeda's fluid agenda.

The third and fourth models of a terrorist organization's end—achievement of the group's cause and transition toward a political role, negotiations, or amnesty—also bear little relevance to al-Qaeda today. It is hard to conceive of al-Qaeda fully achieving its aims, in part because those aims have evolved over time, variably including achievement of a pan-Islamic caliphate, the overthrow of non-Islamic regimes, and the expulsion of all infidels from Muslim countries, not to mention support for the Palestinian cause and the killing of Americans and other so-called infidels. Historically, terrorist groups that have achieved their ends have done so by articulating clear, limited objectives. Al-Qaeda's goals, at least as articulated over recent years, could not be achieved without overturning an international political and economic system characterized by globalization and predominant U.S. power. As the historical record indicates, negotiations or a transition to a legitimate political process requires feasible terms and a sense of stalemate in the struggle. Also, members of terrorist groups seeking negotiations often have an incentive to find a way out of what they consider a losing cause. None of this describes bin Laden's al-Qaeda.

This points to another issue. As al-Qaeda has become a hybrid or "virtual organization," rather than a coherent hierarchical organization, swallowing its propaganda and treating it as a unified whole is a mistake. It is possible that bin Laden and his lieutenants have attempted to cobble together such disparate entities (or those entities have opportunistically attached themselves to al-Qaeda) that they have stretched beyond the point at which their interests can be represented in this movement. Some of the local groups that have recently claimed an association with al-Qaeda have in the past borne more resemblance to ethnonationalist/separatist groups such as the PLO, the IRA, and the LTTE. Examples include local affiliates in Indonesia, Morocco, Tunisia, and Turkey. This is not to argue that these groups' aims are rightful or that their tactics are legitimate. Rather, because of its obsession with the notion of a monolithic al-Qaeda, the United States is glossing over both the extensive local variation within terrorist groups and their different goals; these groups have important points of divergence with al-Qaeda's agenda, and the United States does no one any favors by failing to seriously analyse and exploit those local differences (except perhaps al-Qaeda). The U.S. objective must be to enlarge the movement's internal inconsistencies and differences. Al-Qaeda's aims have become so sweeping that one might wonder whether they genuinely carry within them the achievement of specific local grievances. There is more hope of ending such groups through traditional methods if they are dealt with using traditional tools, even including, on a case-by-case basis, concessions or negotiations with specific local elements that may have negotiable or justifiable terms (albeit pursued through an illegitimate tactic). The key is to emphasize the differences with al-Qaeda's agenda and to drive a wedge between the movement and its recent adherents.

The historical record of other terrorist groups indicates that it is a mistake to treat al-Qaeda as a monolith, to lionize it as if it is an unprecedented phenomenon with all elements equally committed to its aims, for that eliminates a range of proven counterterrorist tools and techniques for ending it. It is also a mistake to nurture al-Qaeda's rallying point, a hatred of Americans and a resentment of U.S. policies (especially in Iraq and between Israel and the Palestinians), thereby conveniently facilitating the glossing over of differences within. Fifth, reducing popular support, both active and passive, is an effective means of hastening the demise of some terrorist groups. This technique has received much attention among critics of George W. Bush and his administration's policies, many of whom argue that concentrating on the "roots" of terrorism is a necessary alternative to the current policy of emphasizing military force.

This is a superficial argument, however, as it can be countered that a "roots" approach is precisely at the heart of current U.S. policy: the promotion of democracy may be seen as an idealistic effort to provide an alternative to populations in the Muslim world, frustrated by corrupt governance, discrimination, unemployment, and stagnation. But the participants in this debate are missing the point. The problem is timing: democratization is a decades-long approach to a short-term and immediate problem whose solution must also be measured in months, not just years. The efforts being undertaken are unlikely to have a rapid enough effect to counter the anger, frustration, and sense of humiliation that characterize passive supporters. As for those who actively sustain the movement through terrorist acts, it is obviously a discouraging development that many recent operatives have lived in, or even been natives of, democratic countries—the March 2004 train attacks in Madrid and 2005 bombings in London being notable examples.

The history of terrorism provides little comfort to those who believe that democratization is a good method to reduce active and passive support for terrorist attacks. There is no evidence that democratization correlates with a reduction in terrorism; in fact, available historical data suggest the opposite. Democratization was arguably the cause of much of the terrorism of the twentieth century. Moreover, democracy in the absence of strong political institutions and civil society could very well bring about radical Islamist governments over which the West would have little influence. There are much worse things than terrorism, and it might continue to be treated as a regrettable but necessary accompaniment to change were it not for two potentially serious developments in the twenty-first century: (1) the use of increasingly destructive weapons that push terrorist attacks well beyond the "nuisance" level, and (2) the growing likelihood that terrorism will lead to future systemic war. In any case, the long-term, idealistic, and otherwise admirable policy of democratization, viewed as "Americanization" in many parts of the world, does not represent a sufficiently targeted response to undercutting popular support for al-Qaeda. There are two vulnerabilities, however, where cutting the links between al-Qaeda and its supporters hold promise: its means of funding and its means of communication. Efforts to cut off funding through traditional avenues have had some results. But not nearly enough attention is being paid to twenty-first-century communications, especially al-Qaeda's presence on the internet. The time has come to recognize that the stateless, anarchical realm of cyberspace requires better tools for monitoring, countermeasures, and, potentially, even control. Previous leaps in cross-border communication such as the telegraph, radio, and telephone engendered counteracting developments in code breaking, monitoring, interception, and wiretapping. This may seem a heretical suggestion for liberal states, especially for a state founded on the right of free speech; however, the international community will inevitably be driven to take countermeasures in response to future attacks. It is time to devote more resources to addressing this problem now. Western analysts have been misguided in focusing on the potential use of the internet for so-called cyberterrorism (i.e., its use in carrying out attacks); the internet is far more dangerous as a tool to shore up and perpetuate the al-Qaeda movement's constituency. Preventing or interdicting al-Qaeda's ability to disseminate its message and draw adherents into its orbit is crucial. Countering its messages in serious ways, not through the outdated and stilted vehicles of government websites and official statements but through sophisticated alternative sites and images attractive to a new generation, is an urgent priority.

As for the opponent's marketing strategy, al-Qaeda and its associates have made serious mistakes of timing, choice of targets, and technique; yet the United States and its allies have done very little to capitalize on them. In particular, the United States tends to act as if al-Qaeda is essentially a static enemy that will react to its actions, but then fails to react effectively and strategically to the movement's missteps.

The Bali attacks, the May 2003 attacks in Saudi Arabia, the Madrid attacks, the July 2005 London attacks—all were immediately and deliberately trumpeted by al-Qaeda associates. Where was the coordinated counterterrorist multimedia response? There is nothing so effective at engendering public revulsion as images of murdered and maimed victims, many of whom resemble family members of would-be recruits, lying on the ground as the result of a terrorist act. Outrage is appropriate. Currently, however, those images are dominated by would-be family members in Iraq, the Palestinian territories, Abu Ghraib, and Guantanamo Bay.

The West is completely outflanked on the airwaves, and its countermeasures are virtually nonexistent on the internet. But as the RIRA, PFLP-GC, and ETA cases demonstrate, the al-Qaeda movement can undermine itself, if it is given help. A large part of this "war" is arguably being fought not on a battlefield but in cyberspace. The time-honoured technique of undermining active and passive support for a terrorist group through in-depth analysis, agile responses to missteps, carefully targeted messages, and cutting-edge technological solutions is a top priority.

This is a crucial moment of opportunity. Polls indicate that many of al-Qaeda's potential constituents have been deeply repulsed by recent attacks. According to the Pew Global Attitudes Project, publics in many predominantly Muslim states increasingly see Islamic extremism as a threat to their own countries, express less support for terrorism, have less confidence in bin Laden, and reflect a declining belief in the usefulness of suicide attacks. In these respects, there is a growing range of commonality in the attitudes of Muslim and non-Muslim publics; yet the United States focuses on itself and does little to nurture cooperation. American public diplomacy is the wrong concept. This is not about the United States and its ideals, values, culture, and image abroad: this is about tapping into a growing international norm against killing innocent civilians—whether, for example, on vacation or on their way to work or school—many of whom are deeply religious and many of whom also happen to be Muslims. If the United States and its allies fail to grasp this concept, to work with local cultures and local people to build on common goals and increase their alienation from this movement, then they will have missed a long-established and promising technique for ending a terrorist group such as al-Qaeda.

There is little to say about the sixth factor, the use of military repression, in ending al-Qaeda. Even though the U.S. military has made important progress in tracking down and killing senior operatives, the movement's ability to evolve has demonstrated the limits of such action, especially when poorly coordinated with other comparatively underfunded approaches and engaged in by a democracy. Although apparently effective, the Turkish government's repression of the PKK and the Peruvian government's suppression of Shining Path, for example, yield few desirable parallels for the current counterterrorist campaign.

Transitioning out of terrorism and toward either criminality or full insurgency is the final, worrisome historical precedent for al-Qaeda. In a sense, the network is already doing both. Efforts to cut off funding through the formal banking system have ironically heightened the incentive and necessity to en-gage in illicit activities, especially narcotics trafficking. With the increasing amount of poppy seed production in Afghanistan, al-Qaeda has a natural pipeline to riches. This process is well under way. As for al-Qaeda becoming a full insurgency, some analysts believe this has already occurred. Certainly to the extent that Abu Musab al-Zarqawi and his associates in Iraq truly represent an arm of the movement (i.e., al-Qaeda in Iraq), that transition is likewise well along.

The alliance negotiated between bin Laden and al-Zarqawi is another example of an effective strategic and public relations move for both parties, giving new life to the al-Qaeda movement at a time when its leaders are clearly on the run and providing legitimacy and fresh recruits for the insurgency in Iraq. As many commentators have observed, Iraq is an ideal focal point and training ground for this putative global insurgency. The glimmer of hope in this scenario, however, is that the foreigners associated with al-Qaeda are not tied to the territory of Iraq in the same way the local population is, and the tensions that will arise between those who want a future for the nascent Iraqi state and those who want a proving ground for a largely alien ideology and virtual organization are likely to increase—especially as the victims of the civil war now unfolding there continue increasingly to be Iraqi civilians. The counter to al-Zarqawi's al-Qaeda in Iraq, as it is for other areas of the world with local al-Qaeda affiliates, is to tap into the long-standing and deep association between peoples and their territory and to exploit the inevitable resentment toward foreign terrorist agendas, while scrupulously ensuring that the United States is not perceived to be part of those agendas. If the United States continues to treat al-Qaeda as if it were utterly unprecedented, as if the decades-long experience with fighting modern terrorism were totally irrelevant, then it will continue to make predictable and avoidable mistakes in responding to this threat. It will also miss important strategic opportunities. That experience points particularly toward dividing new local affiliates from al-Qaeda by understanding and exploiting their differences with the movement, rather than treating the movement as a monolith. It is also crucial to more effectively break the political and logistical connections between the movement and its supporters, reinvigorating time-honoured counterterrorism tactics targeted at al-Qaeda's unique characteristics, including the perpetuation of its message, its funding, and its communications. Al-Qaeda continues to exploit what is essentially a civil war within the Muslim world, attracting alienated Muslims around the globe to its rage-filled movement. Al-Qaeda will end when the West removes itself from the heart of this fight, shores up international norms against terrorism, undermines al-Qaeda's ties with its followers, and begins to exploit the movement's abundant missteps.

AL-QAEDA UNIQUE AMONG TERRORIST ORGANIZATIONS

Four characteristics distinguish al-Qaeda from its predecessors in either nature or degree: its ºuid organization, recruitment methods, funding, and means of communication.

Fluid Organization

The al-Qaeda of September 2001 no longer exists. As a result of the war on terrorism, it has evolved into an increasingly diffuse network of affiliated groups, driven by the worldview that al-Qaeda represents. In deciding in 1996 to be, essentially, a "visible" organization, running training camps and occupying territory in Afghanistan, al-Qaeda may have made an important tactical error; this, in part, explains the immediate success of the U.S.-led coalition's war in Afghanistan. Since then, it has begun to resemble more closely a "global jihad movement," increasingly consisting of web-directed and cyber-linked groups and ad hoc cells. In its evolution, al-Qaeda has demonstrated an unusual resilience and international reach. It has become, in the words of Porter Goss, "only one facet of the threat from a broader Sunni jihadist movement." No previous terrorist organization has exhibited the complexity, agility, and global reach of al-Qaeda, with its ºuid operational style based increasingly on a common mission statement and objectives, rather than on standard operating procedures and an organizational structure.

Al-Qaeda has been the focal point of a hybrid terrorist coalition for some time, with ties to inspired freelancers and other terrorist organizations both old and new. Some observers argue that considering al-Qaeda an organization is misleading; rather it is more like a nebula of independent entities (including loosely associated individuals) that share an ideology and cooperate with each other. The original umbrella group, the International Islamic Front for Jihad against Jews and Crusaders, formed in 1998, included not only al-Qaeda but also groups from Algeria, Bangladesh, Egypt, and Pakistan.

A sampling of groups that are connected in some way includes the Moro Islamic Liberation Front (Philippines), Jemaah Islamiyah (Southeast Asia), Egyptian Islamic Jihad (which merged with al-Qaeda in 2001), al-Ansar Mujahidin (Chechnya), al-Gama'a al-Islamiyya (primarily Egypt, but has a worldwide presence), Abu Sayyaf (Philippines), the Islamic Movement of Uzbekistan, the Salafist Group for Call and Combat (Algeria), and Harakat ul-Mujahidin (Pakistan/Kashmir). Some experts see al-Qaeda's increased reliance on connections to other groups as a sign of weakness; others see it as a worrisome indicator of growing strength, especially with groups that formerly focused on local issues and now display evidence of convergence on al-Qaeda's Salafist, anti-U.S., anti-West agenda.

The nature, size, structure, and reach of the coalition have long been subject to debate. Despite claims of some Western experts, no one knows how many members al-Qaeda has currently or had in the past. U.S. intelligence sources place the number of individuals who underwent training in camps in Afghanistan from 1996 through the fall of 2001 at between 10,000 and 20,000; the figure is inexact, in part, because of disagreement over the total number of such camps and because not all attendees became members. The International Institute for Strategic Studies in 2004 estimated that 2,000 al-Qaeda operatives had been captured or killed and that a pool of 18,000 potential al-Qaeda operatives remained. These numbers can be misleading, however: it would be a mistake to think of al-Qaeda as a conventional force, because even a few trained fighters can mobilize many willing foot soldiers as martyrs.

Methods of Recruitment

The staying power of al-Qaeda is at least in part related to the way the group has perpetuated itself; in many senses, al-Qaeda is closer to a social movement than a terrorist group. Involvement in the movement has come not from pressure by senior al-Qaeda members but mainly from local volunteers competing to win a chance to train or participate in some fashion.

The process eems to be more a matter of "joining" than being recruited, and thus the traditional organizational approach to analyzing this group is misguided. But the draw of al-Qaeda should also not be overstated: in the evolving pattern of associations, attraction to the mission or ideology seems to have been a necessary but not sufficient condition. Exposure to an ideology is not enough, as reflected in the general failure of al-Qaeda to recruit members in Afghanistan and Sudan, where its headquarters were once located. As psychiatrist Marc Sageman illustrates, social bonds, not ideology, apparently play a more important role in al-Qaeda's patterns of global organization.

Sageman's study of established links among identified al-Qaeda operatives indicates that they joined the organization mainly because of ties of kinship and friendship, facilitated by what he calls a "bridging person" or entry point, perpetuated in a series of local clusters in the Maghreb and Southeast Asia, for example. In recent years, operatives have been connected to al-Qaeda and its agenda in an even more informal way, having apparently not gone to camps or had much formal training: examples include those engaged in the London bombings of July 7 and 21, 2005, the Istanbul attacks of November 15 and 20, 2003, and the Casablanca attacks of May 16, 2003.

This loose connectedness is not an accident: bin Laden describes al-Qaeda as "the vanguard of the Muslim nation" and does not claim to exercise command and control over his followers. Although many groups boast of a connection to al-Qaeda's ideology, there are often no logistical trails and thus no links for traditional intelligence methods to examine. This explains, for example, the tremendous difficulty in establishing connections between a radical mosque, bombers, bomb makers, supporters, and al-Qaeda in advance of an attack (not to mention after an attack).

Another concern has been the parallel development of Salatist networks apparently drawing European Muslims into combat against Western forces in Iraq. The European Union's counterterrorism coordinator, Gijs de Vries, for example, has cautioned that these battle-hardened veterans of the Iraq conflict will return to attack Western targets in Europe.

The Ansar al-Islam plot to attack the 2004 NATO summit in Turkey was, according to Turkish sources, developed in part by operatives who had fought in Iraq.

A proportion of those recently drawn to the al-Qaeda movement joined after receiving a Salafist message disseminated over the internet. Such direct messages normally do not pass through the traditional process of vetting by an imam. European counterterrorism officials thus worry about members of an alienated diaspora— sometimes second-and third-generation immigrants— who may be vulnerable to the message because they are not thoroughly trained in fundamental concepts of Islam, are alienated from their parents, and feel isolated in the communities in which they find themselves. The impulse to join the movement arises from a desire to belong to a group in a context where the operative is excluded from, repulsed by, or incapable of successful integration into a Western community.

Thus, with al-Qaeda, the twentieth-century focus on structure and function is neither timely nor sufficient. Tracing the command and control relationships in such a dramatically changing movement is enormously difficult, which makes comparisons with earlier, more traditional terrorist groups harder but by no means impossible; one detects parallels, for example, between al-Qaeda and the global terrorist movements that developed in the late nineteenth century, including anarchist and social revolutionary groups.

Means of Support

Financial support of al-Qaeda has ranged from money channelled through charitable organizations to grants given to local terrorist groups that present promising plans for attacks that serve al-Qaeda's general goals.

The majority of its operations have relied at most on a small amount of seed money provided by the organization, supplemented by operatives engaged in petty crime and fraud. Indeed, beginning in 2003, many terrorism experts agreed that al-Qaeda could best be described as a franchise organization with a marketable "brand."

Relatively little money is required for most al-Qaedaassociated attacks. As the International Institute for Strategic Studies points out, the 2002 Bali bombing cost less than $35,000, the 2000 USS *Cole* operation about $50,000, and the September 11 attacks less than $500,000.

Another element of support has been the many autonomous businesses owned or controlled by al-Qaeda; at one point, bin Laden was reputed to own or control approximately eighty companies around the world.

Many of these legitimately continue to earn a profit, providing a self-sustaining source for the movement. International counterterrorism efforts to control al-Qaeda financing have reaped at least $147 million in frozen assets. Still, cutting the financial lifeline of an agile and low-cost movement that has reportedly amassed billions of dollars and needs few resources to carry out attacks remains a formidable undertaking.

Choking off funds destined for al-Qaeda through regulatory oversight confronts numerous challenges.

Formal banking channels are not necessary for many transfers, which instead can occur through informal channels known as "alternative remittance systems," "informal value transfer systems," "parallel banking," or "underground banking." Examples include the much-discussed *hawala* or *hundi* transfer networks and the Black Market Peso Exchange that operate through family ties or unofficial reciprocal arrangements.

Value can be stored in commodities such as diamonds and gold that are moved through areas with partial or problematical state sovereignty. Al-Qaeda has also used charities to raise and move funds, with a relatively small proportion of gifts being siphoned off for illegitimate purposes, often without the knowledge of donors. Yet efforts to cut off charitable ºows to impoverished areas may harm many genuinely needy recipients and could result in heightened resentment, which in turn may generate additional political support for the movement. Al-Qaeda's fiscal autonomy makes the network more autonomous than its late-twentieth-century state-sponsored predecessors.

Means of Communication

The al-Qaeda movement has successfully used the tools of globalization to enable it to communicate with multiple audiences, including potential new members, new recruits, active supporters, passive sympathizers, neutral observers, enemy governments, and potential victims.

These tools include mobile phones, text messaging, instant messaging, and especially websites, email, blogs, and chat rooms, which can be used for administrative tasks, fund-raising, research, and logistical coordination of attacks. Although al-Qaeda is not the only terrorist group to exploit these means, it is especially adept at doing so.

A crucial facilitator for the perpetuation of the movement is the use of websites both to convey messages, *fatwas,* claims of attacks, and warnings to the American public, as well as to educate future participants, embed instructions to operatives, and rally sympathizers to the cause.

The internet is an important factor in building and perpetuating the image of al-Qaeda and in maintaining the organization's reputation. It provides easy access to the media, which facilitates al-Qaeda's psychological warfare against the West. Indoctrinating and teaching new recruits is facilitated by the internet, notably through the dissemination of al-Qaeda's widely publicized training manual (nicknamed "The Encyclopedia of Jihad") that explains how to organize and run a cell, as well as carry out attacks.

Websites and chat rooms are used to offer practical advice and facilitate the fraternal bonds that are crucial to al-Qaeda. In a sense, members of the movement no longer need to join an organization at all, for the individual can participate with the stroke of a few keys. The debate over the size, structure, and membership of al-Qaeda may be a quaint relic of the twentieth century, displaced by the leveling effects of twenty-first-century technology.

The new means of communication also offer practical advantages. Members of al-Qaeda use the web as a vast source of research and data mining to scope out future attack sites or develop new weapons technology at a low cost and a high level of sophistication. On January 15, 2003, for example, U.S. Secretary of Defence Donald Rumsfeld quoted an al-Qaeda training manual retrieved by American troops in Afghanistan that advised trainees that at least 80 percent of the information needed about the enemy could be collected from open, legal sources.

5

Pakistan's New Generation of Terrorists

INTRODUCTION

Pakistani authorities have long had ties to domestic militant groups that help advance the country's core foreign policy interests, namely in connection with Afghanistan and India. Since Islamabad joined Washington as an ally in the post-9/11 "war on terror," analysts have accused Pakistan's security and intelligence services of playing a "double game," tolerating if not outright aiding militant groups killing NATO troops in Afghanistan. Pakistan denies these charges.

Concerns about Pakistan's commitment to counterterrorism heightened in May 2011, when U.S. commandos killed al-Qaeda mastermind Osama bin Laden at a compound not far from Islamabad. Leadership elements of al-Qaeda and the Afghan Taliban have made Pakistan's semiautonomous tribal areas their home, where they often work with a wide variety of Islamist insurgent groups like the Haqqani Network. Some groups have used Pakistan as a staging ground for attacks in Afghanistan, while others have pursued domestic targets, including schools and houses of worship, as well as organs of the state.

TERRORIST GROUPS

The numerous terrorist groups operating in Pakistan have tended to fall into one of the five categories laid out by

Ashley J. Tellis, a senior associate at Carnegie Endowment for International Peace, in a January 2008 Congressional testimony:

- Sectarian: Religiously motivated groups such as the Sunni Sipah-e-Sahaba and Lashkar-e-Jhangvi and the Shia Tehrik-e-Jafria that are engaged in violence within Pakistan
- Anti-Indian: Groups focused on the Kashmir dispute that operate with the alleged support of the Pakistani military and the intelligence agency Inter-Services Intelligence (ISI), such as Lashkar-e-Taiba, Jaish-e-Muhammad, and Harakat ul-Mujahadeen
- Afghan Taliban: The original Taliban movement and especially its Kandahari leadership centred around Mullah Mohammad Omar, believed to be based in Quetta
- Al-Qaeda and its affiliates: The global jihadist organization founded by Osama bin Laden and led by Ayman al-Zawahiri
- The Pakistani Taliban: A coalition of extremist groups in the Federally Administered Tribal Areas (FATA), led by Mullah Fazlullah

Other militant groups fall outside of Tellis' framework, including secessionist groups such as the Balochistan Liberation Army in southwest Pakistan.

But with greater coordination among groups, experts say, lines have blurred. The Haqqani Network, a semiautonomous faction of the Taliban and a U.S.-designated Foreign Terrorist Organization, is emblematic of the complex interrelations among militant groups in both Pakistan and Afghanistan. A 2011 report from the Combating Terrorism Center (CTC), an independent research institution based at the U.S. Military Academy at West Point, characterizes the group as a "nexus player" with ties to Pakistan's ISI, al-Qaeda, and Uzbek militants, among others. "For the past three decades, the Haqqani Network has functioned as an enabler for other groups and as the fountainhead (*manba*) of local, regional, and global militancy," write Don Rassler and Vahid Brown in the report. The group's leading financier and emissary, Nasiruddin Haqqani, was killed near Islamabad in November 2013 under uncertain circumstances; three other senior leaders were killed in U.S. drone strikes in the two years prior. Sirajuddin Haqqani, Nasiruddin's elder brother, leads the group.

The revelation that Osama bin Laden had been hiding in a compound around the corner from the Pakistan military academy at Kakul raised new questions about the ISI's commitment to counterterrorism.

THE PAKISTANI TALIBAN

Supporters of the Afghan Taliban who sought refuge in Pakistan's tribal areas morphed into a distinct entity following the Pakistani army's initial incursion into the semiautonomous region in 2002. In December 2007, about thirteen disparate militant groups coalesced under the umbrella of Tehrik-i-Taliban Pakistan (TTP), also known as the Pakistani Taliban, led by Baitullah Mehsud of South Waziristan. Pakistani authorities accused him of orchestrating the assassination of former prime minister Benazir Bhutto in December 2007. Short-lived ceasefires signed with Islamabad in 2008 and 2009 provided opportunities for the Pakistani Taliban to regroup and make territorial gains, analysts say.

After a U.S. drone strike killed Baitullah in August 2009, his cousin and deputy Hakimullah Mehsud assumed leadership of the TTP. Hakimullah was reportedly prepared to take part in imminent peace talks with Islamabad when he was killed in a U.S. drone strike along with a top deputy in November 2013. But analysts say the prospects for peace talks were dim. Hakimullah declared war against the state, saying in October 2013: "Pakistan's system is un-Islamic, and we want it replaced with an Islamic system. This demand and this desire will continue even after the American withdrawal [from Afghanistan]." Stephen Tankel, scholar at the Carnegie Endowment for International Peace, notes that if talks had been allowed to fail, Pakistani public opinion would have turned more decisively against the Taliban rather than the United States, which many blame for the insurgency's resilience.

A shura council chose hard-liner Mullah Fazlullah as Hakimullah's successor shortly after his death. Fazlullah, who gained infamy for ordering the assassination attempt on Pakistani schoolgirl and activist Malala Yousafzai, rejected talks with the government. Analysts question whether Fazlullah can maintain TTP cohesion as the first emir from outside the Mehsud tribe.

The predominantly Pashtun group draws membership from all of FATA's seven agencies as well as several settled districts of Khyber Pakhtunkhawa in the northwest. The TTP has declared jihad against the Pakistani state, seeks to control territory, enforces sharia, and fights NATO forces in Afghanistan. "We will target security forces, government installations, political leaders, and police," Asmatullah Shaheen, head of the shura council that selected Mullah Fazlullah, told Reuters, adding, "We will not target civilians, bazaars, or public places. People do not need to be afraid."

It's difficult to assess the size of the Pakistani Taliban. "There are not reliable estimates of how large the TTP is, largely due to challenges associated with even defining the borders of the group and the loose-knit nature of how it is organized along either subtribal or subregional lines," CTC's Rassler says.

The Pakistani Taliban has targeted security forces and civilians alike; among its most audacious attacks have been bombings of Islamabad's Marriott Hotel in September 2008, which killed at least sixty people, and Peshawar's Pearl Continental Hotel in June 2009, in which seventeen were killed. TTP expressed transnational ambitions when it claimed responsibility for a failed bombing in New York's Times Square in May 2010.

The Punjabi Taliban, a loose conglomeration of militant groups of Punjabi origin, gained prominence after major 2008 and 2009 attacks in the cities of Lahore, Islamabad, and Rawalpindi. The network has both sectarian and Kashmir-oriented aims. It has chafed at the Pakistani Taliban's central leadership, *Jane's Intelligence Review* reported in late August 2013, but is uniquely capable of "mount[ing] complex operations in urban environments," particularly in Punjab, Pakistan's most populous and politically significant province.

The Haqqani Network, whose operations straddle the porous Afghan-Pakistani border known as the Durand Line, has proven a valuable ally to the Pakistani Taliban in some of these pursuits. The Haqqanis have not only fought alongside the TTP and Afghan Taliban in Afghanistan, but have also served as influential mediators between the TTP and Islamabad. Pakistan has long been a supporter and beneficiary of the Haqqanis, according to CTC. The network has helped Islamabad manage militant groups in FATA, and provided leverage against India in the struggle over Kashmir. Pakistan sees the Pashtun group, which has been among the most lethal to NATO forces in Afghanistan, as a potential source of leverage after the scheduled withdrawal of coalition troops at the end of 2014.

THE CHANGING FACE OF TERRORISM

Violence in Pakistan has been on the rise, particularly since 2007, as terrorist groups have targeted political leaders, the military and police, tribal leaders, minority Shia, and schools. Though virtually unheard of a decade ago, suicide bombings have become ubiquitous in recent years—a reflection of al-Qaeda's influence, experts say. Three such attacks were documented in 2002 and 2003 combined; at the trend's peak in 2009 there were seventy-six attacks, and there were thirty-seven in the first ten months of 2013, according to the New Delhi–based South Asia Terrorism Portal (SATP).

Besides providing militant groups in Pakistan with technical expertise and capabilities, al-Qaeda also promotes cooperation among them. CTC's Rassler wrote in 2009 that al-Qaeda "assumed a role as mediator and coalition builder among various Pakistani militant group factions by promoting the unification of entities that have opposed one another or had conflicting ideas about whether to target the Pakistani state."

The Taliban, meanwhile, has become ever more entrenched in Pakistan, building a nationwide network by finding common cause with terrorist groups that target the Shia and the Pakistani state while establishing roots—and a lucrative criminal enterprise—in Karachi. Pakistani paramilitary Rangers launched a campaign in September 2013 to address the city's criminal and terrorist groups, reportedly arresting over 1,500 suspectsin a month. Meanwhile, Pakistan's political parties advocated negotiations with the Taliban in part to stave off even higher levels of violence in Punjab and other populated areas, Tankel writes. SATP reported 2,745 civilians and 601 security forces killed in terrorist violence in the first ten months of 2013—roughly on pace with the prior two years.

"Ample evidence exists of tacit Pakistani consent and active cooperation with the drone program." —International Crisis Group

Counterterrorism Challenges

Pakistani security forces have at times struggled to muster the capacity and will to confront domestic militants, even though the army and police are increasingly targeted by militant groups. Some experts say that since the bloody encounter between Pakistan's security forces and militant Islamic students in Islamabad's Red Mosque in 2007, some groups previously under state patronage broke away. In October 2009, militants attacked army headquarters in Rawalpindi and held around forty people hostage for over twenty hours. Such attacks heralded a new period in army and ISI relations with many of these militant groups, analysts say.

Even though the Pakistani army and the ISI have been more willing to go after militants, analysts say they continue to form alliances with groups such as the Haqqanis that they can use as a strategic hedge against India and Afghanistan. In a September 2011congressional testimony, then chairman of the Joint Chiefs of Staff Admiral Mike Mullen referred to the Haqqani network as a "strategic arm of Pakistan's Inter-Services Intelligence agency." Pakistan's security establishment has denied these charges.

The revelation in May 2011 that Osama bin Laden had been hiding in a compound around the corner from the Pakistan military academy at Kakul raised new questions about the ISI's commitment to counterterrorism. CIA director Leon Panetta said the agency ruled out partnering with Pakistan on the bin Laden mission out of concern that it would be compromised. President Asif Ali Zardari, writing in the immediate aftermath of the operation, said allegations that Pakistan harboured terrorists amounted to "baseless speculation."

The CIA has conducted an extensive targeted killing campaign to supplement Pakistani counterterrorism efforts, particularly in the rugged, remote terrain of North and South Waziristan. U.S. drones are currently launched from Afghan soil, but it's unclear whetherthis arrangement will continue after the scheduled U.S. withdrawal in 2014. If the targeted killing program is called off, veteran intelligence analyst Bruce Riedel argues, "Al-Qaeda will regenerate rapidly in Pakistan. Its allies like the Taliban and Lashkar-e-Taiba will help it to rebuild. The ISI will either turn a blind eye or, worse, a helping hand." The program's detractors have questioned the United States' ability to distinguish between militants and civilians, and argue that strikes may contribute to radicalization in the frontier provinces.

Prime Minister Nawaz Sharif, who began an unprecedented third term in June 2013, has railed against U.S. drone strikes as an affront to Pakistani sovereignty while advocating for talks with the TTP. Yet, the International Crisis Group notes, "Ample evidence exists of tacit Pakistani consent and active cooperation with the drone program." Pakistan's leadership seeks greater say over targeting, the ICG says, "often to punish enemies, but sometimes, allegedly, to protect militants" with whom the security services have cooperative relations—including elements of the Haqqani Network and Taliban.

Meanwhile, Pakistan's own counterterrorism efforts have come under the scrutiny of human rights observers. Amnesty International and Human Rights Watch charged security forces with torture, extrajudicial killings, arbitrary detention of tribal-area residents, and the enforced disappearance of "journalists, human rights activists, and alleged members of separatist and nationalist groups." A new legal framework awaiting Parliament's approval would relieve an overwhelmed criminal justice system by establishing new federal courts equipped to handle terrorism cases, advocates say. But critics caution that the law would codify prolonged, warrantless detentions by the state.

ISI AND ITS CHICANERY IN EXPORTING TERRORISM

Spying is the second oldest profession and nowhere else has it flourished more in a short span of time, that too, with professional élan, than in Pakistan. Known by an innocuous sounding Inter Services Intelligence, abbreviated ISI, it is a wellspring of power and one of the most virile intelligence agencies in the Third World.

Its forte lies in intrigue, hatching conspiracies, brokering terrorism and paddling misinformation.

In recent months, the ISI has been in the limelight, its skulduggery having received a new thrust; - and notoriety, fresh showing. A report has been submitted to the House of Representatives by a US task force on terrorism and unconventional warfare. The report, titled, *"The New Islamic International"* is significant in revealing the exploits of the ISI. That the original document is dated 1 Feb, 1993 and that it has some inaccuracies, does not make the findings any less weighty.

A chilling account" of the role of the ISI's involvement with Sikh militants, bombing of *Kanishka* and American, indifference, even complicity, has been chronicled in a TV documentary of the "Fifth Estate" of the Canadian Broadcasting Corporation (CBC). The documentary deplores the stark incertitude of the US administration, although it had all the evidence at its command. What is worse, by overplaying India's human rights record, Washington has rewarded and encouraged ISI in its evil pursuits. Another TV feature produced by CNN, traces the ISI connection to Afghan *Mujahideen* and their involvement with the bomb explosion at the World Trade Centre. The two documentaries together, and the Canadian one by itself constitute the most damning indictment of the ISI Involvement in terrorism in India.

There is no let-up in ISI's activities in Kashmir, which include recruitment, training, arming and induction of terrorist bands, besides guidance in planning and conduct of operations. ISI has a strong connection with Sheikh Mubarak Shah Jiiani and his fundamentalist outfit AI Fuqra. The group was responsible for "a decade long string of assassinations and bombings in the name of Islamic purity."' Lately, the ISI has ventured in the heartland through promotion of narco- terrorism chauvinism. That the organisation has a collaborator in the Nepalese Parliament, is sinister enough, but far more portentous is the patron-age bestowed by the Bangladesh Government to the organisation for promoting insurgency in the North East. Recently, a diabolic plot has been unravelled, which suggests that ISI, with a view to destabilising the country and creating mayhem, had planned to repeat the

Bombay blasts and ignite communal riots all over India to coincide with the Republic Day celebrations. With the arrest of four Pakistanis, one Bangladeshi and one Indian in the capital by the Delhi police, a major breakthrough in exposing diabolic designs of the ISI has been achieved. "Though intelligence agencies are still trying to fully comprehend the ramifications of the latest ring unearthed by them, the case is perhaps the most brazen attempt of Pakistan-trained nationals being sent to create terror countrywide; the low intensity proxy war transcending to practically open hostilities."

MISSION, BUDGET AND ORGANISATION

The ISI was founded in 1948 by a British army officer, Maj Gen R Cawthome, then Deputy Chief of Staff in Pakistan Army. He conceived it as part of the military establishment, intended to combine and co-ordinate intelligence set-ups of the Services. Over the years, it gathered influence and when Zia seized power, it acquired real muscle. Today, the ISI has achieved the dubious distinction of being the most dreaded outfit, within; and a master-hand in dispensing terrorism, without.

ISI charter incorporates gathering of external and internal intelligence; co-ordination of intelligence functions of the three Services; surveillance over its cadre, foreigners, media men, politically conscious segments of Pak society, diplomats of other countries accredited to Pakistan and Pak diplomats serving outside the country; interception and monitoring of communications; and conduct of covert offensive operations.

Unlike intelligence agencies of other countries, the top and middle rung officers of the ISI are exclusively drawn from the military establishment. Its chief is designated as director general and appointed from amongst the serving lieutenant generals; although there has been one exception when a retired officer was assigned to the post. During Yahya Khan's rule, the DG also headed the newly created National Security Council and that added to his stature and influence. Under the DG there are three deputy director generals (DDGs), one each from the army, the navy and the air force. The ISI mans a Military Liaison Section (MLS) in the Ministry of Interior. ISI is a major beneficiary of Pakistan's national budget, with a large unaccountable chunk coming from the defence outlay. In Pakistan, no one knows, not even the Prime Minister, as to how much ISI costs to run or precisely how many people it employs. But today, the ISI enjoys total support of the Prime Minister.

The fact that, both, she and her father had suffered at the hands of ISI has been forgotten and treated as a closed chapter. She has learnt the bitter lesson from her previous tenure as the Prime Minister, when she confronted the organisation and tried to clip its wings. In the battle of wits, the ISI won and retained its clout; she lost and was sent packing. ISI continues to call the shots, the inspired views of its devaluation notwithstanding.

During the late 80s the most powerful component of the organisation was Joint Intelligence Bureau (JIB), which handled political intelligence. In her earlier incarnation as the PM, Benazir ordered transfer of the functions of JIB to the Interior Ministry, but the rumour has it that the sensitive files and dossiers were, instead, moved to the GHQ. Later, they found their way back to the very drawers and cabinets from which they were taken out. And now the ISI has become a law unto itself.

An equally powerful component of the ISI is the Joint Counter Intelligence Bureau (JCIB), it continues to wear authority and influence on its sleeves.

It has a director for field surveillance, who keeps a watch on Pak diplomats accredited to other countries. The bureau conducts intelligence operations in Asia and the Middle East. Its special spheres of interest are. the countries of South Asia, a common knowledge; and China, a fact not so well known. ISI's operations in China were started in collaboration with the CIA and perhaps continue to be so; this activity is passionately disguised and denied for obvious reasons. Lately, Afghanistan and Muslim republics of the former Soviet Union too, are being handled by JCIB and these countries have become a favourite haunt of ISI agents." The wing also keeps IS Directorate under surveillance through a section referred to as ISSS and prepares reports for the chief executive; for the latter purpose, there is a director attached to the PM's Secretariat.

Joint Intelligence Miscellaneous (JIM) deals with espionage in foreign countries and offensive intelligence operations. J&K affairs, including infiltration, exfilteration, propaganda and shady operations, is the mission assigned to JIN (an abbreviation standing for Joint Intelligence : North) under DDG External II, a post generally held by a major general. While the IN is staffed by military technocrats, who could be described in modern parlance as system analysts, the cloak and dagger stuff is the preserve of operational cells and forces, dedicated to the mission. Those exclusively set up for J&K have been periodically wound up and then re-created with the aim of causing confusion. During Zia's time, a special cell was established for Afghanistan which had under it deputy directors responsible for political affairs, training, arms distribution and refugees. This was disbanded after the Jallalabad fiasco. Although no precise information is available, it is believed that a similar cell was created for imparting training and distribution of arms to the Sikh extremists and it continues to be active.

Another field in which the ISI has been highly successful, is gathering of signal Intelligence. This mission is assigned to JSIB (Joint Signal Intelligence Bureau) which has DDs (Deputy Directors) Wireless, Monitoring and Photos on its establishment. The organisation runs a chain of intercept stations along the entire Indo-Pak border, besides providing communication support to the militants operating in the Valley. In the spring of 1992, it was estimated that there were about 200 clandestine radio stations operating on the Indian soil.

JIX is the largest wing, which serves as the secretariat. It co-ordinates the functions of other wings and field organisations, prepares intelligence estimates and threat perceptions, besides giving administrative support to the organisation.

ZIA'S CONTRIBUTION TO ISI'S GROWTH

It is tragic that in the middle ages, most Islamic politics degenerated and settled down as personal despotism, the egalitarian spirit and appeal of the religion notwithstanding. Aggressive Islamization in Pakistan, true to historical legacy, has proceeded in tandem with Zia's totalitarian statecraft. Zia found in the ISI a useful instrument to synthesise the two with a view to eliminating internal opposition and promoting fanaticism, which *inter alia* included such insidious ploys as exporting terrorism and conducting raids in the name of *jihad.* He cleverly manipulated the diverse, yet coterminous aphorism of Khomeini's Islamic revolution and American foreign policy objectives in support of policy of subversion in Afghanistan and when that showed promise, made a grand design to destabilise India.

A strong component of American foreign policy interest at the fag end of the Cold War was to ostracise communist influence in Afghanistan and teach the former Soviet Union a lesson for its interventionist policies.

"In pursuance of this objective CIA forged a close relationship with the ISI, through which three billion dollars worth of arms were channelled to the Afghan *Mujahideen*. Hekmatyar was adopted as the intelligence agencies favourite surrogate to prosecute America's Proxy war against the Soviet Union." Zia utilised the opportunity not only to rehabilitate his tarnished international image by becoming a front-line warrior against the communist menace but also to settle scores with India.

During Zia's reign, the ISI fuelled the separatist movements in Punjab and Kashmir. In domestic affairs, it acquired a special status and immense power. Experience in Afghanistan, where in collusion with the CIA, It conducted one of the biggest covert operations in the world since the end of the Vietnam war gave its teeth a sharper bite. It luxuriated in joint collaboration with CIA and was privy to arcane ways, learning the latest tricks of the trade. In Afghanistan, it played one group of the *Mujahideen* against the other and then presided over the assemblage to compose differences and broker accords. It did not spare its mentors, the CIA. Without their knowledge and concurrence, it helped itself to dollars and siphoned off arms meant for *Mujahideen* to Iran, an arch enemy of the US. When the beans were spilled and the US decided to send a fact-finding and stock-taking board, it fudged records, put Ojhri camp to the torch and destroyed the incriminating evidence.

Internal Involvements

In his book, *If I am Assassinated,* Bhutto charged that the ISI was actively used to spy on him, whereas it had failed miserably when it came to gathering hard intelligence in both, 1965 and 1971 operations. He writes, "How did Ayub Khan and Yahya Khan use the intelligence agencies? Yahya Khan used to the hilt the intelligence agencies for political purposes to divide the politicians and influence the elections of 1979." In the same vein, he repeats, "Ayub Khan also used the intelligence agencies for political purposes to the hilt... He tried to prevent my party from getting off the ground." Narrating the episode when Ayub Khan sought an explanation about the inability of the ISI to locate the Indian Armoured Division, he quotes DG, ISI Brig Riaz Hussain of having replied, "Sir, from June 1964, Military Intelligence has been given political assignments on elections and post-election repercussions."

Bhutto on ISI

Bhutto, in the book *If I am Assassinated* writes extracts of the White Paper issued by the Zia administration that it "demonstrates its piety with crocodile tears on the role of the intelligence agencies of the State as a political arm of the government of Pakistan People's Party." He writes that on page 195, the White Papers registers it concern in the following words.. "The role of the intelligence agencies of the State as a political arm of the PPP regime, particularly in relation to the general elections, raises many disconcerting questions. When politics permeates such sensitive institutions as the Intelligence Bureau or the ISI, it naturally deflects them from their prime concern with the State's external and internal security. Political bias against dissenting political parties which are a very necessary component of a democratic society, also tends to complicate and distort the task of State security."

Bhutto refutes the allegations and writes that Lt Gen G Jilanl was DG of ISI before he became President of Pakistan on 20 Dec 1971, that except for him, "all the officials incharge of intelligence at the federal level were arrested on the night of the coup, or within a month of it," that Jilanl was "not touched but on the contrary was sent to the Defence Ministry as its Secretary." He further adds, "This question must be considered in conjunction with Lt Gen Jilanl's successful effort in influencing me to consider the then Maj Gen Zia-ui-Haq for the post of the Chief of Staff in suppression of about six Generals. *This is only a fraction of the story.* But even with this *minimum disclosure* I would like to ask who exploited whom? Did the Military Intelligence Chief and his Chief of Staff exploit me or I exploited them?".

Undoubtedly Bhutto suffered at the hands of the ISI, but he cannot be absolved of his contribution to the breeding of this Frankenstein. Of his own admission, he mooted the suggestion for "the merger of Central Intelligence Agencies into one integrated intelligence department divided into two categories (i) internal and (ii) external. Obviously, he conceived a greater role for the ISI, hoping to keep his position secure by placating the military's most powerful army, but the chicks came home to roost. It was Zuifiqar, who legalised the ISI's involvement in domestic surveillance, for which he paid heavily. Benazir, too, had wrecked her reputation and suffered a smash-up when she tried to take on the ISI. Today, she is an ardent supporter of the organisation and a collaborator of its misdemeanours.

Proxy War in Kashmir

The extent of Pakistani and Afghan influence on the Islamic transformation of the Kashmiri insurgency is quite clear. In mid-Eighties, Islamic revivalism had taken a "radical political stance", slogans advocating establishment of an Islamic state were publicly raised and received with growing popularity. The population came under the influence of the leadership of Jamaati-islami and Khomeinists. By 1984, an Islamic radicalisation had developed that saw the rise of such movements as JKLF, Kashmir Liberation Front, the Mahaz-e-Azadi and the Liberation League.

By 1985, Jamaat-e-lsiami and Al-Jihad movements had become influential in Kashmir politics. Islamic indoctrination was provided by Jamaat-isiami of Pakistan". This is the same organisation about whose leaders Bhutto had written that, "they show a total ignorance of Islam and betray an un-islamic mentality", they "follow the ways of *jahiliyat* (imbecility)" and accused them of denouncing *Qaid-i-Azam* as*Kafir-e-Azam.*

Al Jihad took inspiration from the ideology of the Iranian revolution. It publicly pronounced that "Islamic revolution" was the only way to liberate Kashmir. In a short span of a few years, "there was a marked erosion of the secular Kashmiri personality and a Muslim identity with fundamentalist overtones started emerging rapidly." It lent justification to give the movement "a pan-Islamic character and an extra territorial dimension."

In the early stages, the ISI used *Mujahideen* infrastructure to help the Kashmiri and Sikh separatists. "At times the assistance was funnelled through Afghan rebel leader Gulbuddin Hekmatyar's Hizb-i-isiami group, thus providing 1siamabad with deniability."

The report submitted by the task force to the American Congress mentions creation of a special force, which made its debut in July 1991 and within months perpetuated communal frenzy and accelerated militancy. This was the time when Pak sponsored terrorism was at its peak. The novel feature of the special force is that it has been drawn from those with operational experience in Afghanistan. Officered by mainly Punjabis, the cadre is highly motivated and willing to undertake daring assignments. They have let loose agents provocateur, who freely mingle with government functionaries in the Valley and have been known to use Indian army and police uniforms. Recent events show that they have been effective in creating bad blood between the two arms of the security outfit, pitching one against the other. A hot ploy is to commit atrocities the garb of the security forces, so that the finger of accusation is pointed in the wrong direction.

The report points out that about 20,000 young Kashmiris have been trained and armed in POK in recent years. It dwells at length over the Pakistani involvement in stoking the fire of terrorism and insurgency in Kashmir. "Indeed the very size of the Pakistani training programme is telling," it observes; and goes on to describe the breakdown of the number of trainees in the ISI run camps and their cabal affiliation.

In the manner of Maoist classical approach, 1987 to 1989 marked the first stage of insurgency. It aimed at seeding discontent and creating a nucleus of Islamic militancy; and was characterised by sporadic uncoordinated attacks on soft targets. The insurgents adopted hit and run tactics and avoided direct confrontation with the security forces. "Only expendable, barely trained terrorists were committed." Had the Indian authorities dealt with the insurgents more firmly and nipped the secessionist movement in the bud, the ISI would have suffered a dismal defeat, but the Government dilly-dallied; worst, it openly aired differences within its own ranks on the approach to be adopted. This led to the second stage, more professionalized and better co-ordinated. The assassination of Mirwaiz provided the right kind of opportunity for garnering active support for the terrorists.

In early 1992, the ISI had established a common command over the disparate military arms of organisations that had mushroomed, true to the example set by the *Mujahideen* of Afghanistan and the general pattern of Islamic militancy elsewhere. It succeeded in the fall of 1991, in mediating and settling an agreement between the military arms of the Hizb-ul- Mujahideen, the Allah Tigers and the Ikhwan-ul-Musalmeen to launch joint and co-ordinated operations. Though somewhat tenuous, the ISI control helped in funnelling arms, ammunition and money to the militants, besides conducting training and indoctrination programmes. The training campus had started turning out more hardened and motivated gangs, well-versed in the use of sophisticated weapons, explosives and radio sets. Whereas a total of 390 cases of terrorism were reported in 1988, the number spurted to 4,971 in 1992. There was a substantial increase in the incidents against the security forces from 6 in 1988 to a high of 3,413 in 1992. The quality and the quantity of arms captured by the security forces, too, is indicative of the growing involution of the ISI. In 1988, only 34 AK-47s (or its later versions), were recovered; the figure went up to a whopping 3,775 in 1992.

The second stage was also marked by the setting up of a number of broadcasting stations in POK. These spewed communal venom and created a "spiral of hatred and violence between the security forces and the masses." Gremlin broadcasts and mischievous propaganda led to influencing the gullible and the devout, who came out on the streets with increased frequency and virulence. Sada-i-Hurriet became the instrument for hatching and mongering rumours. The fare that it dished out in local languages, is the fertile product of ISI propaganda mill. A notable achievement of the ISI is the influencing of the foreign media like AFP and BBC, whose misrepresentation of happenings in India has seriously affected peace and serenity of minorities and led to escalation of tension in the Valley.

EXPORTING TERRORISM THROUGH SIKH MILITANCY

Pakistan, since its creation, has never been well disposed towards Sikhs. Pakistani writings project Sikhs as barbarians, heap ridicule and pass blasphemous and disparaging remarks about the universally venerated *gurus*. In early Eighties, the tactics changed though not the attitude. *Operation Blue Star* provided the long awaited opportunity, which the ISI exploited thoroughly in creating communal fracas in Punjab and fuelling Sikh community's alienation with the government. ISI drew a crafty game-plan, initially designed to supply arms and giving refuge to those who committed crimes in India. Later, the scope was enlarged and an operation codenamed K-2 launched, which *inter-alia* included training to the estranged Sikh youth in the use of sophisticated arms and explosives, co-ordination with militants operating in the Valley and directing terrorist acts, both, in India and abroad. Terrorism received a fresh fillip with Indira Gandhi's assassination. As for Islamabad, this was a positive "proof of the strategic value of subversion."

By 1985 the ISI had established a vast training infrastructure for the Afghan resistance movement that could "just as well be used for training and support of other regional groups." Terrorists of Dal Khalsa were chosen for importing advance training in Afghan *Mujahideen* camps. A few of these trainees were killed in a Soviet raid on an Afghan training camp in Pakistan and highly incriminating documents recovered from them. It was not long before CIA trained Afghan terrorists, too, were inducted into India with the purpose of organising acts of terrorism against members of the Indian government and foreign diplomatic representatives.

There was a strategic motive too, highlighted by the US task group's report to the Congress. Pakistan's claims to Kashmir tempted ISI to sponsor and encourage creation of Khalistan. In its calculation, this would make "the Indian defence of Kashmir difficult."" Islamabad was determined to exploit growing tension in Kashmir to destabilise India and, therefore, embarked on an ambitious plan of providing training and military assistance to Punjab militants."

What could be more incriminating evidence of Pak involvement in Punjab than Benazir's own admission that she had Helped Rajiv Gandhi's government in overcoming Sikh militancy? She said this in an interview with the BBC, broadcast on 13 Feb, 1994. Howsoever, she and spokesmen of her government thed to explain away the substance of the interview, the import of this "confession", as Nawaz Shahef described her *faux pas,* is not lost on the international audience.

ISI'S ACTIVITIES IN THE NEIGHBOURING COUNTRIES

Elsewhere in India and other countries of South Asia, ISI is no less active. It has resurrected its old contacts, which it had assiduously cultivated in erstwhile East Pakistan and which collaborated with Pak occupation forces during the Bangladesh struggle for independence. At a press conference in Shillong at the conclusion of the 37th meeting of the North Eastem Council, the Home Minister accused Dhaka of providing a base to the ISI for its operations. Although Bangladesh has feigned ignorance and denied the charge, it appears from Mr Chawan's categorical statement, that his ministry has incontrovertible evidence of Bangladeshi complicity in promoting the activities of the ISI to disrupt peace in insurgency hit north east. He observed that, "it is unfortunate that even some officers in uniform have imparled training to the militants of the region to create large-scale law and order problem." His warning that unless Bangladesh government desists from supporting ISI, it could "create problems between the two countries," is timely and may invoke salutary effect.

The Republic Day plot was detected just in time by the Indian counter intelligence. The trail led to Dilshed Mirza Beg, a Member of Parliament in Nepal, belonging to Sadbhavana party. He has acted as a conduit for supplying weapons and explosives to a network of agents, spread all over India. Some links and moles have been identified; arrests too have been made, but this may be only a tip of the iceberg. Recent explosion in Odeon cinema in the capital appears to be the handiwork of ISI agents. Whereas Delhi, Secunderabad, Bombay, Lucknow and towns in J&K and Punjab are kept under watch, it is well nigh impossible to cover the entire country. The ISI has both resources and design to create mischief anywhere in India.

Anti-India Lobbying in Britain and the US

Britain, Canada and the US have become important centres for lobbying and fund raising for anti-India activities, organised by the ISI. In Britain there has been a sudden spurt in the number of Muslim charities, not dedicated to purely humanitarian aims. *The Guardian* named two of these organisations, The "Young Muslims" and "The Islamic Foundation". The latter is headed by Prof. Khurshid Ahmed who is vice president of Jammat-i-isiami which has promoted extremism in Afghanistan and Kashmir.

Khalistan lobby in the US is no less active. The case of Dr. Badett is a cogent example. On 11 Feb he along with 28 Congressmen, initiated an appeal to the President, urging him to pressure the Indian government into letting Amnesty International investigate alleged human rights abuses in Punjab. The letter to Clinton was released by the Council of Khalistan. It accused the Indian government of brutal repression against the Sikh nation. The signatories say "We are concemed at the bloodshed in Punjab, Khalistan and the human rights organisations such as Amnesty International, be permitted to investigate human lights violations in the Sikh homeland."

Last year, Bariett had opposed an anti-India resolution promoted by the well known India basher Mr Den Burton. Badett's turnaround is attributed to active courting by the Khalistani lobby, exhorted by the ISI. The Khalistanis have promised hefty contributions to Bariett's election campaign against Neil Dhillon a democratic candidate, in the battle of hustings for the Maryland Constituency distinct constituency.

Concluding Remarks

The US may turn a Nelson's eye to the macabre doings of the ISI and for political expediency show reluctance to declare Pakistan a terrorist state, but the bitter truth rancours. "US sources support the assertion that Washington was well aware of Pakistan's involvement with militant Sikhs during the 80s but were reluctant to make an issue of it for fear of jeopardising the campaign to drive the Soviets out of Afghanistan. Canadian intelligence sources confirm that neither the Federal Bureau of Investigation nor the CIA provided any useful assistance during the investigation of *Kanishka* bombing. Prosecutors in World Trade Centre case argue that the conspiracy was actually hatched in Peshawar.

What happened in Afghanistan is being repeated in Kashmir. There are as many "Azaadi" groups as was the motley that constituted Mujahideen. God forbid if they too achieve their objective, the fate of Kashmir will be no different from that of Afghanistan. Pakistan will exploit every opportunity to rip the secular and multiethnic fabric of the country. Their capacity to fish in the troubled waters of the north east as ISI has done in Kashmir, is an ominous warning. But the seeds of discord can only sprout where there is social inequity and political indifference. The government cannot depend merely on letters of protests to Pakistan or the meagrely funded counter-intelligence to meet the challenge posed by the ISI. It will be highly desirable to involve the people and build their resistance to exploitation. For that, it is necessary that genuine grievances of the venerable section of the society are removed and elements that have been alienated, brought back to the fold.

6

Understanding Pakistan's Approach to the War on Terror

INTRODUCTION

Although Pakistan has been a frontline state in the war on terror since the tragic events of September 11, 2001, there is no doubt that General Musharraf initially cast his lot with the United States mainly as a result of deep fears about what U.S. enmity might imply for Pakistan's long-standing rivalry with India, its efforts at economic revival, its nuclear weapons program, and its equities in the conflict over Kashmir.

Desirous of protecting Islamabad's interests in these areas and to avoid Pakistan becoming a target in the campaign against terrorism, Musharraf reluctantly cut loose Islamabad's ties with the Taliban— a force it had nurtured, trained, and equipped for almost a decade in its effort to secure control over Afghanistan—and stood aside as the U.S.-led coalition assisted its detested antagonist, the Northern Alliance, to rout its own clients and their al- Qaeda accomplices and seize power in Kabul. Because the al-Qaeda elements in Afghanistan during the 1990s were never directly dependent on the Pakistani Inter-Services Intelligence Directorate (ISID) for their success (despite maintaining a significant liaison relationship), the ejection of their Arab, African, and Central Asian mercenaries was viewed with fewer misgivings than the flight of the Pashtun- dominated Taliban, who were tied to Pakistan directly in terms of both patronage and ethnicity.

Islamabad's ties to the Taliban were so strong and so important that throughout the initial phase of Operation Enduring Freedom, General Musharraf and his cohort implored the United States to desist from decisively destroying Mullah Muhammad Omar's regime in Afghanistan. When this objective could not be secured, Pakistani leaders argued against all coalition military operations that would result in ejecting the Taliban's foot soldiers from their traditional bases in the southeastern provinces of the country. When these entreaties were also disregarded by the United States and the comprehensive defeat of the Taliban appeared inevitable as a result of joint U.S. and Northern Alliance military operations, Islamabad responded by covertly exfiltrating its army and intelligence personnel seconded to the movement— along with some key Taliban operatives, if Indian intelligence sources are accurate— while permitting the defeated stragglers to cross over to safety across the frontier and into Pakistani territory. Taliban forces and their al- Qaeda guests thus found their way across the highly porous Afghanistan- Pakistan border into the FATA. While Pakistani border patrols concentrated their efforts against the latter group, resulting in the seizure of numerous low- level al- Qaeda elements, these operations nonetheless were never rigorous or watertight enough and were, in any event, frustrated by other factors: the old tribal tradition of extending succor to strangers who ask for protection; the region's history of providing foot soldiers, first, for the anti- Soviet jihad and, later, for the war against the Northern Alliance in Afghanistan; the absence of a strong Pakistani state presence in this area; and the utterly hostile topography consisting of remote and difficult mountain terrain with poor lines of communication, all of which combined to bestow on the defeated remnants a substantial measure of sanctuary and assistance.

The U.S. inauguration of the "global war on terror" soon compelled General Musharraf to make good on his "principled" decision to join the U.S.-led coalition. This inevitably required Musharraf to confront the sources of terrorism that had developed internally in Pakistan, most of which ironically resulted from his own army's previous decisions to nurture radical Islamist organizations because of their utility to Islamabad's military campaigns in Kashmir and Afghanistan.

Four different terrorist groupings were implicated in this regard. The first were the domestic sectarian groups like the Sunni Sipah- e- Sahaba and its offshoot the Lashkar- e- Jhangvi and the Shia Tehrik- e- Jafria Pakistan and its offshoot the Sipah- e-Muhammad, which were engaged in violent bouts of bloodletting within the country. Although many of these groups had enjoyed the support of the Pakistani government, the military, and the intelligence services previously, their unexpected growth in power over time had become not only an embarrassment to their sponsors but also a serious challenge to domestic order. As Christine Fair has summarized it, "The scale of sectarian violence in Pakistan is staggering, with hundreds of people killed or injured in such attacks each year." The New Delhi–based Institute for Conflict Management has documented sectarian violence alone as claiming close to 5,000 lives in Pakistan since 1989, with incidents involving everything from targeted killings of high-profile civilians, to bombings of mosques and drive- by shootings of innocents, to pitched gun battles in major population centers. In one incident, for example, sectarian hostility in the town of Parachinar in the Kurram Agency involved a five- day war, where small arms, mortars, rocket launchers, and antiaircraft missiles were all used in a convulsive spasm that claimed hundreds of lives and injured many more.

Confronted by such challenges to the writ of his state, Musharraf was only too happy to exploit the opportunities offered by the war on terror to crack down on these groups and suppress them once and for all. He did so, however, only selectively. Focusing the government's energies primarily on those Deobandi and Shia groups whose objectives were out of sync with the military's perception of the national interest, he targeted Sunni groups such as the Lashkar- e- Jhangvi, the Sipah- e- Sahaba Pakistan, the Harkat- ul- Mujahideen al- Alami, the Jundullah, and to a much lesser degree the Harkat- ul- Jihad- e- Islami, as well as Shia threats such as the Sipah- e- Muhammad, primarily because they were engaged in "anti- national" jihadi violence within Pakistan rather than in support of Islamabad's external ambitions vis-à- vis India and Afghanistan. Using the entire panoply of coercive state capabilities, these entities were therefore put down with a heavy hand through arrests, targeted assassinations, and aggravated intergroup massacres. Although many of the tools used to defeat these perpetrators of sectarian violence were often unconstitutional, Musharraf shrewdly judged that the aggressive dismemberment of these groups would not evoke either domestic or international condemnation. He was right.

THE GLOBAL WAR ON TERROR

Using the opportunities therefore afforded by the global war on terror, the Pakistani security services systematically eliminated many sources of sectarian violence within two years of the campaign's initiation, even though they have been unable to conclusively eradicate the cancer of sectarian bloodshed within Pakistan itself. In part, this is due to the selectivity of Musharraf's antisectarian campaign. But the continuing fragmentation of these violent groups; their links to the wider networks of international terrorism now resident in Pakistan, various foreign sponsors abroad, and the flourishing madaris within the country; and the continuing utility of their gun- toting membership to different political parties and occasionally to governmental organs themselves imply that sectarian threats will be impossible to extinguish so long as "state policies of Islamisation and [the] marginalization of secular democratic forces" continue to persist in Pakistan.

The second set of groups, the terrorist outfits operating with Pakistan Army and ISID support against India in Kashmir, was treated in a remarkably different way compared with the anti-national sectarian militants inside Pakistan. These groups, such as the Lashkar- e- Toiba, the Jaish- e- Muhammad, and the Harkat-ul- Mujahideen, for example, were the long lances in the Pakistani campaign to wrest the disputed state of Jammu and Kashmir from India. Since the late 1980s, the Pakistani military has financed, trained, armed, and launched these cadres on their murderous missions into Kashmir and elsewhere inside the Indian union. Because the struggle for control over the disputed Himalayan state was fully under way by the time the global war on terror was inaugurated, these terrorist groups were more or less exempted from Musharraf's domestic campaign against violence and extremism.

This exclusion was justified both on the strategic rationale that Pakistan's participation in the war on terror was intended, among other things, to protect its freedom of action in Kashmir and on the repeated, though fraudulent, assertion that these groups, far from being terrorists, only personified the legitimate Kashmiri struggle for self- determination against India. In fact, of all the Pakistani- sponsored Deobandi terrorist groups operating against India in Kashmir and elsewhere, only one entity— the Hizbul Mujahideen— began life as an indigenous Kashmiri insurgent group; the others, including the most violent organizations such as the Lashkar- e- Toiba, the Jaish- e- Muhammad, and the Harkat- ul- Mujahideen, are all led, manned, and financed by native Pakistanis. This reality notwithstanding— and perhaps because of it— Islamabad continued to sustain the operations of these groups against India but, in an effort to maintain the consistency of its commitments to the global war on terror, now began to emphasize that its support took only the form of moral, and not material, encouragement.

This charade was rudely interrupted by the December 13, 2001, terrorist attack on India's parliament when, in response to New Delhi's subsequent military mobilization, Pakistan was compelled by U.S. diplomacy to initiate a series of measures to restrict the activities of its terrorist clients. The implementation of these actions, however, was at best halfhearted and inconsistent.

Far from seeking to extirpate these terrorist groups permanently, Musharraf sought mainly to defang India's threats of military action and to alleviate Washington's fears of an inconvenient Indo- Pakistani war. His overarching objective consisted of protecting these terrorist assets to the extent possible because they represented national investments— a "strategic reserve"—in Islamabad's subconventional war against New Delhi.

Consequently, to this day, Musharraf has not sought to eliminate the Deobandi terrorist groups operating against India in Kashmir and elsewhere; he has instead sought only to modulate their activities, depending on the extent of satisfaction he derives from the prevailing state of diplomatic relations with New Delhi and the progress secured in the ongoing Indo- Pakistani peace process.

Although the Pakistani- supported infiltration of terrorist groups into Kashmir— but not into the rest of India— appears to have abated in recent years, most observers conclude that this phenomenon is linked either to Musharraf's desire not to provide India with any excuses to abandon the generally fragile peace process or to domestic crises within Pakistan. In any event, it is agreed that Musharraf simply has not made the decisive decision to abandon or eliminate the terrorist groups operating against India in the manner witnessed, for example, in the case of the more virulent anti- national sectarian entities operating within Pakistan.

The third group relevant to the Pakistani decision to join the U.S.-led coalition against terrorism consisted of the Taliban, that is, the Pashtun remnants of the regime ejected from power in Kabul as a result of the initial success of Operation Enduring Freedom. After their defeat at the hands of the Northern Alliance, the Taliban cadres hastily returned to the regions whence they originated.

Many in the rank and file withdrew to their villages in the southern Afghan provinces of Helmand, Kandahar, Oruzgan, and Zabol as well as along the border areas on the western side of the Durand Line separating Afghanistan from Pakistan, that is, in the eastern Afghan provinces of Paktika, Paktia, Khowst, Nangarhar, and Konar. Given their significance as high- value targets, the core Taliban leadership— along with those Pakistani Pashtuns who had joined their movement— crossed over the Afghanistan-Pakistan border into the relative safety of the FATA.

Because most of the Taliban's fighters originally mobilized by the Pakistani ISID were drawn from the Ghilzai confederation of Pashtuns, which dominates eastern and southeastern Afghanistan, and from the other Pashtun tribes inhabiting the FATA, their return to these ancestral lands was not surprising. In fact, all the evidence relating to the incidence of terrorist attacks since 2001 suggests strongly that the war- fighting cadres of the Taliban continue to remain bivouacked in these areas.

The exact location of the supreme leadership of the Taliban movement, however, cannot be established with any self- evident clarity. Irrespective of where the rahbari shura (leadership council) centred on Mullah Omar and his closest associates found shelter in the immediate aftermath of their defeat, Afghan military and civilian intelligence officials as well as NATO commanders today believe that this coterie eventually found refuge in Quetta, the largest city and capital of Pakistan's Baluchistan Province, from where they continue to operate to this day.

As Col. Chris Vernon, NATO's chief of staff for southern Afghanistan, declared forthrightly, "The thinking piece of the Taliban [operates] out of Quetta in Pakistan. It's the major headquarters— they use it to run a series of networks in Afghanistan."

These networks, in turn, are judged to be directed by four subsidiary shuras based in Quetta, Miran Shah, Peshawar, and Karachi: the first three actually control or coordinate most of the ongoing terrorist operations occurring, respectively, along the southern, central, and northern "fronts" in Afghanistan, whereas the fourth is believed to connect the Taliban with the logistics, financial, and technical assistance conduits emanating from the wider Islamic world. The pattern of terrorist attacks occurring in Afghanistan from 2002 to 2007, again corroborates this judgment.

Because the Pakistani state was most intimately involved in the creation of the Taliban before their fall, Musharraf's antiterrorism campaign after September 11, 2001, deliberately avoided any concerted targeting of this group and, in particular, its senior leadership. No other explanation is consistent with the fact that, although Pakistani military, intelligence, and paramilitary forces apprehended scores of al- Qaeda operatives, including numerous key individuals in the al- Qaeda hierarchy, the senior Taliban leaders killed or captured in southern Afghanistan or in the FATA have numbered literally a handful in comparison.

This asymmetry in seizures is all the more odd because, prior to Operation Enduring Freedom, Pakistani military and ISID liaison elements were deeply intertwined with all levels of the Taliban mand structure and its war fighters in the field. In contrast, the Pakistani intelligence relationship with al- Qaeda in Afghanistan was more tenuous, yet Pakistan's military forces were able to apprehend far more al- Qaeda cadres than Taliban operatives.

These successes in regard to al- Qaeda have invariably been attributed by Pakistanis, including General Musharraf, to the fact that it was always easier to identify the ethnically alien al- Qaeda elements along the frontier in comparison with the Taliban who, being ethnically Pashtun, were able to disguise their identities by assimilating into the larger tribal population.

While this explanation is only partly true— non- native fighters have lived in and become amalgamated into the social structures of the Afghan- Pakistani frontier since at least the anti-Soviet jihad of the 1980s— it is also disingenuous because the Pakistani ISID was not only deeply involved in the recruitment, training, arming, and operations of Taliban fighters at multiple bureaucratic levels, but it also maintained an intense liaison relationship with the Ghilzai tribes whose population has been disproportionately represented in the Taliban. Since protecting these relationships was deemed to be especially critical for Pakistan's national security interests in the aftermath of the Northern Alliance victory in Kabul, the large number of Taliban foot soldiers who made their way into the FATA were largely ignored by Pakistani counterterrorism operations so long as they did not engage in any untoward activities that either called attention to their presence or magnified the troubles confronting the Pakistani state.

All told, then, the Taliban network, just like the Pakistani-aided terrorist groups operating in Kashmir and elsewhere in India, was deliberately permitted to escape the wrath of General Musharraf's counterterrorism operations in the initial phase of the war on terror.

Such an approach, however, could not be extended to the fourth group, al- Qaeda, which had also taken up sanctuary in the FATA, particularly in South Waziristan initially. Although al- Qaeda continued to have sympathizers within the extreme fringes of Pakistani society even after the terrible events of September 11 were conclusively attributed to its operations, the Pakistani military establishment did not enjoy the luxury of slackening its campaign against this target because of the consequences for U.S.-Pakistani relations at a time when bilateral ties were just recovering after a decade of U.S. disfavor and when Washington had just embarked on a ferocious campaign against al- Qaeda worldwide.

Most senior Pakistani military officers at the corps command level were also genuinely horrified by the destruction that al-Qaeda wreaked in New York and Washington and, fearing for their country's own future in the face of the monster now present in their midst, supported Musharraf's decision to engage and destroy this terrorist organization of global reach.

Pakistan's military, accordingly, began to prosecute the war against al- Qaeda with great vigor, if not always with finesse, through multiple instruments. These included providing the United States and its military with facilities and access for the prosecution of Operation Enduring Freedom in Afghanistan, and conducting various law enforcement and internal security operations (sometimes in cooperation with their U.S. counterparts) aimed at interdicting terrorist financing and apprehending and rendering terrorist targets for prosecution abroad.

Most important, however, the Pakistani military initiated Operation Al Mizan, a large- scale effort that involved moving major military formations from the Army's XI Corps and elite Special Services Group (SSG) battalions into the FATA, an area where regular army units had not ordinarily been deployed for decades.

These infantry forces joined the Frontier Corps regiments— the paramilitary formations usually located in the region— as a show of force in order to both reassert the strong state presence that historically was lacking and apprehend the al- Qaeda elements that had taken shelter within the area.

This military campaign, which took the form of a gigantic cordon- and- search operation, had several consequences. First, it resulted in the capture of numerous al- Qaeda and other extremist operatives— some 700 at last count— who have since been turned over to the United States.

Because these individuals are mostly foreigners— non–South Asian arrivals living in what are essentially Pashtun lands— detecting their presence, while not easy because of the local support they receive from the natives for ideological reasons and sometimes simply out of greed or fear, was certainly easier.

Second, it forced some though by no means all senior al-Qaeda operatives— for example, Khalid Sheikh Muhammad and Ramzi Binalshibh— to leave the relatively secure FATA sanctuary and disperse further inward into Pakistan, where their insertion into less ideologically congenial surroundings and their need to rely on more complex means of communication increased their susceptibility to detection and arrest.

Third, the dramatic irruption of the Pakistani state into the FATA, through a significant military presence of the kind not seen in more than a century, resulted in making conditions sufficiently inhospitable for al- Qaeda such that its senior leadership and cadres were compelled to relocate under fire from South to North Waziristan and beyond, where they operate to this day.

This forced displacement, which unfortunately remains at continuous risk of reversal, nonetheless had the beneficial effect of disrupting many planned terrorist operations, but the dispersal of the organization's leadership in the northern FATA, especially in the Bajaur Agency where the terrain is inhospitable, the population is violently pro- Taliban, and the presence of the Pakistan Army is thin, has inadvertently made the task of destroying the al- Qaeda core all the more difficult.

In any event, these outcomes suggest that although Pakistan began as a reluctant entrant into the global war on terrorism, it has since become an active participant in the struggle. More than 85,000 Pakistani troops remain garrisoned along the Afghanistan-Pakistan border— a deployment that predates the initiation of the global war on terror.

A significant fraction of these forces, however, is engaged today in counterterrorism operations in the border areas, and more than 600 soldiers have already sacrificed their lives in this effort.

Further, Islamabad itself has now become a victim of terrorism as a variety of groups, ranging from those previously nurtured and now discarded by the Pakistani state, such as the al-Alami faction of the Harkat- ul- Mujahideen, to more distant beneficiaries of past Pakistani policies, such as al- Qaeda, seek to wreak an orgy of revenge against institutions and individuals whom they had previously counted among their sponsors and friends.

That Pakistan has made significant contributions to defeating various terrorist groups is therefore undeniable, yet its larger campaign against terrorism has also been conspicuously selective and perhaps self- serving. While it has secured major gains in eradicating some domestic anti- national sectarian terrorist groups and has contributed disproportionately to the ongoing campaign against al- Qaeda, it has been much more reluctant to conclusively eliminate those terrorist entities operating against India in Kashmir and elsewhere and against Afghanistan both in the FATA and in transit back and forth to the southern and eastern Afghan provinces. Further, the protection of the terrorist infrastructure that supports these groups has produced undesirable blowback because the actors traditionally involved in perpetrating terrorism in Kashmir increasingly either coordinate with or directly assist the Taliban and al- Qaeda in operations against not only Afghanistan but also the United States and even Pakistan itself.

Clearly, strategic and geopolitical calculations play an important part in accounting for this segmented Pakistani response. Islamabad, for example, has long viewed the terrorist groups operating in Kashmir and now in other Indian states as useful instruments for executing its policy of "strategic diversion" against New Delhi. For this reason, Pakistan has been reluctant to target and eliminate these groups conclusively, preferring instead to alternately tighten and loosen control over their operations depending on how much satisfaction it receives from India at any given moment.

The decision to avoid targeting the Taliban was born of similar calculations. Initially, it was owed simply to the inclinations of senior Pakistani military commanders who were just not prepared to add insult to injury by physically eliminating the very forces they had long invested in, especially because they had now suffered the ignominy of having to consent to their client's defeat.

Over time, however, the reasons for protecting the Taliban only grew stronger: India's growing prominence in Afghan reconstruction, its increased influence and presence in Afghanistan more generally, the weakening of the Hamid Karzai government in Kabul, the progressive souring of Pakistani- Afghan relations (including those between Karzai and Musharraf personally), and the disquiet about a possible U.S. exit from Afghanistan (a prospect inferred from the mid-2005 announcement that the United States would divest full command of Afghan combat operations to NATO) once again increased Pakistan's paranoia about the prospect of a hostile western frontier.

It was exactly the desire to avert this outcome that led to the initial Pakistani decision to invest in sustaining the Taliban. And with fear of the wheel turning full circle gaining strength in Islamabad since at least 2005, the temptation to hedge against potentially unfavorable outcomes in Kabul— by protecting the Taliban as some sort of a "force- in- being"—only appeared more and more attractive and reasonable to Pakistan.

Although Pakistan's discriminative approach to fighting terrorism was shaped and implemented by General Musharraf in his dual capacity as president and previously chief of army staff, it would be erroneous to conclude that this prevailing strategy is owed simply to the whim of one man.

This is particularly relevant today when Musharraf's hold on power has become progressively weaker and the future of his political status and effectiveness increasingly clouded. Rather, Musharraf's decisions in regard to counterterrorism strategy since 2001, although publicly perceived as personal dicta, invariably reflected the consensus among the corps commanders of the Pakistan Army and, hence, represent the preferences of Pakistan's military- dominated state.

In other words, even if Musharraf were to suddenly exit the Pakistani political scene at some point, Islamabad's currently disconsonant counterterrorism strategy would still survive so long as the men on horseback continue to be the principal guardians of national security policy making in Islamabad. Because it is unreasonable to expect that the uniformed military will give up its privileges in this regard anytime soon— even if a civilian regime were to return to the helm in the future— the internally segmented counterterrorism policy currently pursued by Pakistan will likely persist for some time to come.

Even if it could be imagined that a civilian dispensation could wrest some control of Pakistan's national security policy from the military, it is not at all certain that the current strategic direction would change dramatically: a civilian regime would probably have greater incentives to combat all sectarian terrorist groups more evenhandedly, but that too is uncertain. After all, both the Pakistan People's Party led by the late Benazir Bhutto and the Pakistan Muslim League led by Nawaz Sharif have had problematic Islamist political allies in the past and, depending on the political exigencies of the moment, could harbor incentives to give even sectarian or otherwise radical entities a breather from prosecution, although that would likely be justified as only a temporary expedient.

Both civilian parties historically also permitted the Pakistani military and intelligence services to aid, abet, and arm the terrorist groups operating in Kashmir and elsewhere in India, sometimes because they were simply powerless to prevent it but at other times with their full knowledge and consent.

It also ought not to be forgotten that even a radically atavistic Islamist group such as the Taliban was raised, promoted, and unleashed by the civilian government of the late Benazir Bhutto (during her second term in office from 1993 to 1996) with the full collaboration of the Pakistani military and intelligence services— and that the Taliban continued to receive complete moral and material support under her civilian successor, the then prime minister, Nawaz Sharif.

CIVILIAN POLITICAL ALTERNATIVES IN PAKISTAN

Both the principal civilian political alternatives in Pakistan would likely continue to prosecute the current antiterrorist operations against al- Qaeda because there seems to be a fragile consensus among Pakistani political elites that this group remains a grave threat to both their country and the international community.

This fact, however, only underscores the continuity that is likely to persist in Pakistan's approach to counterterrorism even if a civilian government were to ascend to power in Islamabad. Although there are likely to be differences in style, nuance, and emphasis, the weaknesses of Pakistan's moderate political parties, Islamabad's enduring interests vis- à- vis Afghanistan and India, and the likely inability of any civilian government to exercise comprehensive control over the Pakistani military and intelligence services all combine to suggest that dramatic changes in attitude and performance toward the Taliban and the terrorist groups operating on Indian soil may not be forthcoming.

And, although sectarian groups within Pakistan may be pursued more uniformly and hopefully just as resolutely as the war against al- Qaeda, the net deviation from Musharraf's currently segmented antiterrorism policies may be either too subtle or too insignificant to really matter.

Ironically, the Bush administration itself bears some responsibility for reinforcing Musharraf's original instincts and entrenching what has now become the enduring Pakistani calculus.

Although President Bush affirmed in the aftermath of the September 11 attacks that his war on terrorism would be total and that states supporting terrorist groups would be required to divest themselves of these entanglements decisively or face America's wrath, his own government never implemented his stirring vision in regard to Pakistan. Rather, during the Indo- Pakistani crisis of 2001–2002—a key moment of truth for Pakistan and its future course in the war on terror— successive U.S. intermediaries visiting the subcontinent pursued an approach that only permitted Islamabad to conclude that the war on terrorism was in fact eminently "divisible." By not pressing Pakistan to relinquish all its terrorist clients once and for all during that crisis— as Washington had previously compelled Islamabad to forsake the Taliban on September 13, 200136—the United States lost a momentous opportunity to help Pakistan rid itself of its long addiction to terror. Instead, the administration's diplomacy, by declining to hold Musharraf accountable for breaching his serial promises to end Pakistani support for terrorism, enabled Islamabad to infer that so long as operatives belonging to "terrorist groups of global reach"—meaning al- Qaeda— were being regularly apprehended by Pakistan, the ISID's links to, and protection of, other regional terrorist organizations would not become a critical liability in U.S.-Pakistan relations.

The liberties thus afforded Pakistan in regard to sustaining its ties with local Kashmiri terrorist groups during the initial phase of the global war on terror consequently reinforced Pakistan's inclination to treat the Taliban remnants similarly.

This blunder had few consequences as long as the Taliban movement was in remission, but it has proved to be a most costly lapse on the part of the United States because the sanctuary afforded by Pakistan to the Taliban— and especially its leadership— since 2002 has only permitted the group to rejuvenate and, once again, to begin offensive operations in Afghanistan that, in effect, threaten to undo the gains secured by the early victories in Operation Enduring Freedom.

The U.S. neglect of the early Pakistani decision to ignore the Taliban as a target of counterterrorism operations can be explained only by the administration's single- minded concentration on the war with al- Qaeda.

This obsession was no doubt justified at the time, but its inadvertent consequences have now come back to haunt the United States, NATO, Afghanistan, and the ongoing military operations associated with Operation Enduring Freedom more generally.

By failing to recognize that the early immunity provided to the Taliban would eventually complicate the effort to defeat al-Qaeda— if for no other reason than that these two groups remain geographically commingled and because Taliban endurance in southern and eastern Afghanistan and in the FATA is an essential precondition for al- Qaeda's survival— the administration lost an opportunity to consolidate its political and military gains in Afghanistan while simultaneously compelling Pakistan to hasten its march away from extremism.

The Bush administration has now begun to press Musharraf to actively interdict the Taliban— an issue that did not become the subject of high- level U.S. demarches before 2005–2006—but it is not certain that, even if responsive Pakistani counterterrorism actions were to be mounted today, they would be as effective as they could have been had they been pursued in the administration's first term. This is because Pakistan's own current intelligence capabilities with respect to the Taliban are probably not as strong as they were when Mullah Omar and his associates were first ejected from Kabul.

Although it is certain that Pakistani information about the Taliban and their leadership is still better than that possessed by other intelligence agencies, including those of the United States, the probable atrophy of Islamabad's connections during the past several years of the war on terror, the strong and growing antagonism within Pakistan toward Musharraf's counterterrorism policies in the FATA, and the increasing opposition from Pakistan's fundamentalist political parties and their social bases of support toward Musharraf's domestic and foreign policies all viciously interact to increase the risk that belated Pakistani actions against the Taliban, including its leadership, may end up being far less successful than they otherwise might have been if executed a few years earlier. And that, in turn, implies not only that the challenges of defeating the Taliban are from a historical perspective rooted in fateful U.S. decisions to treat the Kashmiri terrorists differently when the administration should have done otherwise, but also that Washington ignored the Taliban until it was too late.

CATHOLIC TERRORISTS AND ISLAMISM

There is a genuine issue of interpretation here. It has become a commonplace among analysts of Al Qaeda and of other Middle Eastern terrorist groups to distinguish between what was identified above as "militant Islam" or "Islamism" and the religion practiced by Muslims. This distinction is based not on political correctness but on empirical evidence. Most Muslims, of course, are neither Islamist fundamentalists nor terrorists; many terrorists, however, proclaim they are Muslims—indeed, many proclaim they are the only true Muslims. It is certainly legitimate for Muslims to object to the term *Islamic terrorists* when no one calls the IRA "Christian terrorists" or "Catholic terrorists." It is also unquestionably true that the complaints should be directed less at analysts or journalists who merely report the self-interpretation of others than at the terrorists themselves for their abuse of Islam. So far as most Westerners were concerned, the killing on the morning of September 11 had nothing to do with God; the killers flew out of nowhere and acted in a manner that was utterly irrational, not to say unintelligible. This is why there were so many simplified attempts, in the fall of 2001, to answer a simple question: Why do they hate us?

During the twentieth century, only fourteen terrorist attacks killed more than one hundred people and none killed more than five hundred. Prior to September 11,2001, about a thousand Americans had been killed by terrorists. The sheer magnitude of the killing on that day, when in the course of an hour and a half over three times as many people were murdered, was itself extraordinary. Estimates of property and collateral economic damage are likewise enormous. Including market-based estimates of lowered profits and higher discounts for economic volatility, one source has calculated a price on the order of $2 trillion. However measured, the attacks were unparalleled, and reason enough to understand why the United States has taken steps to reconfigure the architecture of its national security.

It is easy to be overwhelmed by the sheer quantitative size indicated by such data and to limit one's understanding to the entirely intelligible and initial response by military force in Afghanistan and in other places. Moreover, if one does not begin from the experience of shock at the terrorist attacks of September 11,2001, and anger at the killing of so many otherwise innocent individuals, then an initial and immediate understanding of the obvious meaning and significance of those events will have been lost. That is, the spectacular nature of the attacks and the outrage the spectacle was designed to evoke is the proper starting point for a more systematic enquiry into the genesis and meaning of those events. For most ordinary Americans and, indeed, for most ordinary human beings, the attacks of September 11 were a reminder of the difference between good and evil. Indeed, in the context of postmodern liberalism, they were a forceful reminder that there is such a difference. The events of that day were and are not completely open to interpretation and judgments were and are not completely relative or simply matters of opinion. When President Bush referred to the terrorists of September 11 he called them "evil-doers." This was not intended to be rhetorical overkill but an accurate description that can be translated directly into Arabic as *mufsidoon.*

At the same time, however, it is insufficient for political science to register only shock and anger. The indignation of a citizen and the demand for justice or retribution is intelligible enough: most people know who the bad guys are and why they are bad. It is as close to self-evident as a thing can be that the terrorists were and remain religious fanatics and murderers. That is where everyone, both citizens and political scientists, must begin.

There are other problems as well. Some aspects of September 11 were merely technical: simultaneous terrorist attacks, even using old-fashioned car bombs, are rare. The bombings of the American embassies in Nairobi and Dar-es-Salaam on August 7,1998, the eighth anniversary of the arrival of U.S. troops in Saudi Arabia, showed the same skill at synchronizing the attacks, as did the ten-car bomb attack on Bombay in March 1993. Likewise in October 1983, Hezbollah managed to kill 240 people at the U.S. Marine barracks in Beirut along with 60 French paratroopers in the same morning; in 1981 three Venezuelan passenger planes were hijacked, and in 1970 four airliners were taken over by the Popular Front for the Liberation of Palestine, two of which were later destroyed on the ground. The relatively small number of such spectacles raises the question: How did they do it? This question suggests others: How were other such attacks averted in the past? How can they be averted in the future? More than careful planning and coordination were needed to carry out such a spectacular operation: the hijackers were also willing to die, not by risking their lives but with absolute certainty they would not survive their mission. As Hoffman observed:

This dimension of terrorist operations, however, arguably remains poorly understood. In no aspect of the September 11 attacks is this clearer than in the debate over whether all 19 of the hijackers knew they were on a suicide mission or whether only the 4 persons actually flying the aircraft into their targets did. It is a debate that underscores the poverty of our understanding of bin Laden, terrorism motivated by a religious imperative in particular, and the concept of martyrdom. One of the purposes of the present analysis is to provide some clarity on the nature of a suicide mission. That is, in addition to considering such relatively straightforward issues as who did what and how it was accomplished, we are also concerned with interpreting the phenomena of terrorist action as meaningful within the context of sources provided by the terrorists themselves. One of the oldest insights of political science is that all political action is self-interpretive; terrorism is no exception.

There are, moreover, analogies, if not precedents, from within Western political history that provide some degree of guidance. There is, to be sure, an extensive and systematic treatment by specialists on, for instance, Japanese "new religions," on various kinds of liberation movements, and on Islamist sects or the "Christian Identity" movement. There is no shortage of information, although a certain amount of imaginative interpretation is needed to bring coherence to an otherwise complex and fragmentary narrative.

To cast this information into a theoretical context while not forgetting the essential atrocity of terrorism is more difficult. Some guidance through this emotional, intellectual, conceptual, and experiential thicket can be found in an exchange during the early 1950s between Hannah Arendt and Eric Voegelin regarding the appropriate way to study, to come to terms with, and indeed to resist, the novel political reality of ideologically inspired totalitarianism. Notwithstanding their differences, which can easily be exaggerated, both these political scientists began from the direct experience of totalitarian domination and were fully aware that it presented the face of evil in its day. Voegelin, for example, characterized Nazi and Bolshevik revolutions in terms of the "putrefaction of Western civilization" and the "earthwide expansion of Western foulness." Arendt likened the extermination camps to hell on earth. Neither, however, was content simply to record adverse judgment on two varieties of a murderous regime. The last edition of Arendt's book on totalitarianism had grown to over five hundred pages; and even then, after a thorough treatment of the problem, she had serious doubts concerning the wisdom of publishing her analysis because her work might contribute to the continued existence of a phenomenon she wished to destroy and remove from the face of the earth. One might argue as well that Voegelin's books of the late 1930s, his monumental *History of Political Ideas,* written during the 1940s, and, indeed, his classic *New Science of Politics* (1952), were aimed at understanding the origin and significance of the foulness and putrefaction he so obviously deplored.

In April 1938, for example, a month after the Nazi Reich had absorbed Austria, Voegelin published *The Political Religions.* He indicated that the purpose of the book was to analyse and comprehend the self-interpretation of the Nazi movement. In particular, he wished to understand the Nazi claim that the party (and thus its self-interpretation) was the "truth" of the German *Vok* and of the place in history of the reality symbolized as that *Volk.* At the same time, however, he sought to analyse and to draw the connection to the obvious conflict with commonsense reality posed by Nazi race doctrines, associated ideological symbols such as *Volk,* and their oppressive and murderous practices. In other words, in 1938 Voegelin was interested not simply in denouncing the Nazis as "merely a morally inferior, dumb, barbaric, contemptible matter" but in understanding them as "a force, and a very attractive force at that." To put it bluntly, it was a question of comprehending the attractiveness of evil. Evil is no less attractive today. Perhaps more to the point, its attractiveness cannot fully be understood apart from its evilness.

Many conventional analyses of contemporary terrorist acts, as noted above, have described a "religious" dimension among the motivations of terrorist violence. During the 1970s and 1980s, terrorists operating in Northern Ireland used Protestant and

Catholic Christianity as a screen behind which they pursued their own political agendas. More recently, however, larger numbers of terrorist groups "are using religion itself as the primary motivation behind their attacks."

The term *religion,* however, is used in many senses, even in political analyses, and opens up a very wide spatial and temporal field for terrorist activity For example, "ethno-religious" terrorists often direct their violence at "ethno-secular" leaders of the same "religious" group. President Anwar Sadat, for example, was murdered by an Egyptian Islamist and Prime Minister Yitzhak Rabin was murdered by an equivalent type of Israeli. The terrorist group Hamas is as opposed to PLO leader Yasser Arafat as it is to Israel. On occasion divisions between terrorists have led to grotesque contrasts, as when Arafat, responding to Abu Nidal's attempt to assassinate him, said, "He's a real terrorist!" An appropriate comment on this farce was made by Christopher Harmon: "The new terrorists use the old terrorists' arguments. There may be an international interest in preserving Arafat's life, but others must make the moral argument for him; Arafat has no credibility to make it himself." This kind of ambiguity is not confined to the Middle East but applies equally to the Balkans or to the Tamils of South Asia. Sometimes analysts argue that "religious" terrorists are inherently more unpredictable than secular ones because, unlike those of the latter, the objectives of "religious" terrorists are often unintelligible to those who do not share their religious outlook. These and other similarly commonsensical observations express the obvious insight that religion is more than the practices carried on by institutional churches, temples, mosques, and so on, just as politics is more than the practices carried on by states. As Roger Scruton noted at the start of his study of terrorism, the etymology of religion indicates, somewhat ambiguously, the sense of binding together by binding back to an originary event or even a divine revelation. In one way or another, all political orders, including those of the West, are integrated and justified by symbolic narratives that connect political practices in the pragmatic and even secular sense to a larger order of meaning. Thus it is impossible to understand contemporary terrorism without paying close attention to the religiosity or spirituality that terrorists experience as central to their own activities. This can hardly be said to be a new approach in modern political science, to say nothing of the approach adopted by the great philosophers of the past.

For example, about a year before Voegelin's book on political religions appeared, the Japanese ministry of education issued a document announcing the news that Japan was a divine nation. The Japanese, it said, were "intrinsically quite different from so-called citizens of Western nations" because the unbroken bloodlines of its people preserved the "pure" and "unclouded" Japanese spirit. A few years later, in the spring of 1942, several Japanese philosophers assembled in Kyoto to discuss how Japan might "overcome modern civilization," the great embodiment of which had been attacked a few months earlier at Pearl Harbor.

A second example appeared the same year Voegelin's book was published. A Hungarian refugee, Aurel Kolnai, wrote *The War against the West,* an analysis not just of the Nazis but of the Japanese as well. This "war," he said, in 1938, was against the Western notion of citizenship, the state, and the political community. Underlying what Buruma and Mar-galit recently called "occidentalism" are aspects of Western culture that its opponents hate: it is urban, bourgeois, prosperous, and egalitarian. Typically these attributes of Western civilization are denounced as decadent, arrogant, weak, and depraved. Such themes and topics were commonplace during the 1930s. They have reemerged in the early twenty-first century in a modified, but recognizably equivalent, symbolic language, though they are sustained by a quite different spiritual milieu. Hatred of the city, symbolized as Babylon, is considered necessary to ensure the purity and virtue of rural peasant piety. In both the Bible (Gen. 11:4–6) and the Koran (16:23) God takes deep offense at the famous urban monument, the Tower. The contemporary icon of the mythic tower of Babylon is surely the skyline of Manhattan. The anonymity offered by cities and the liberty that such anonymity fosters is seen by the pure as a source of licentiousness and hypocrisy. The separation of public and private, of the word and the heart, is to such individuals the mark of corruption.

Cities are also venues of markets, which the pure detest as being expressions of greed, selfishness, and the cultural decay that comes when human beings, both natives and foreign or immigrant minorities, meet as equals to exchange goods and services. When the Japanese attacked the United States in 1941 they sought, not to take market share away from Americans and Europeans, but to destroy the competitive markets the foreigners created. The Greater East Asia Co-Prosperity Sphere was never intended to be a market. The attack on September 11 was also a reprise of these common themes that reject the city and the market, the settled bourgeois, the petty clerk, or the plump banker, just the sort of person who might have held down a job at the World Trade Center. They are utterly unheroic specimens, the very antithesis of greatness, risk, peril, and sacrifice that so inspires the pure. Moreover, even to traditional Muslims (to say nothing of Islamist terrorists) the modern city is an unwelcome addition to the world. In the words of Abu-Rabi, "The ancient Muslim city, with the mosque, the *madrasah,* and the bazaar at its center, no longer performs a useful function in the eyes of modern capitalism. Far from being sacred and stable, space is subject to continuous change." After Mohammed Atta left Egypt to study urban planning in Germany, it is significant that he wrote a thesis on restoration of Aleppo, an ancient city where the minarets dominated the skyline.

In addition, cities in the West have been the places where laws are drawn up. Courts and jurists, not gods or God, are the sources of modern Western laws. The notion of "man-made law" is not a term of abuse in the contemporary West, in part because of the lessons learned by Westerners following the European religious wars that attended and followed the Reformation of the sixteenth century. The end of Christendom with the settlements at Augsburg (1555) and Westphalia (1698) led to what Westerners now know as freedom of conscience. Moreover, such a liberty requires, in principle, both a secular and a territorially limited government, legitimated not by obedience to God or God's law but by the consent of the governed. Moreover, Westerners have learned from experience that laws made by the spiritually pure in the name of God invariably turn out to be the univocal, undebated decrees of human beings, which Westerners have come to understand to be an attribute of tyranny. Worse, the human beings who rule in the name of God invariably if not inevitably turn out to be males, holy men, with an agenda of oppression, including the oppression of women as women. Among the issues we must consider are the grounds upon which contemporary terrorists justify their attacks on the modern West and, reciprocally, whether the analysis made here of those grounds is merely the application of Western prejudice and unsubstantiated opinion. That is, we aim to provide an analysis that is more than the rationalization of the nevertheless intelligible emotions of an outraged citizen. We may begin in a summary way by recalling the ambivalence, the division, or the tension in Western political science between the city and its citizens and the political scientist or the philosopher, a tension that finds expression in a number of different ways. It can be found most famously in the opening scene of Plato's *Republic*—indeed in its opening word, *kateben.* "I went down," said Socrates, to the port of Athens, Piraeus, the center of its prosperous trade, the market basis of its power, and the symbol of its corruption by that same wealth and power. Moreover, Socrates went down "yesterday," and much as he wished to return to the upper city, the old city, the city of yesterday, he was constrained to stay below, at least for a time. There, with his friends, he used his talents to persuade them to follow him upward and forward to the polis of the Idea, built in their souls and built of speech, after a pattern in heaven.

The insights of Socrates did not exempt him as a citizen from suffering the consequences of the foolish policies of the politicians whose measure he so unfailingly took. As with subsequent expressions of the limits to politics, such as in the Stoic distinction between law and justice or between the two cities of Augustine or the two swords of Pope Gelasius, the language of ambivalence heightens rather than abolishes the distance between the aspiration of the philosopher toward justice or beauty and the pragmatic condition of citizens who must also suffer their absence.

Moreover, one finds in the Bible an equivalent expression of this ambivalence. The Ten Commandments, for instance, were no more God's legislation for the Israelites than Plato's *Republic* was some sort of political "utopia" awaiting establishment by some new and almost divine nomothete. A central biblical distinction (Exodus 20–23) is between the words *(debharim)* of God and God's decisions or ordinances *(mish-patim)*. There is, accordingly, a tension between the two, corresponding approximately to the Ciceronian distinction between the law and justice. Thus, for example, the word of God says (Exod. 20:15) "Thou shalt not steal," but the ordinance of God indicates a legal rule (Exod. 22:1): "if a man steal an ox or a sheep, and kill it, or sell it, he shall pay five oxen for an ox, and four sheep for a sheep," which indicates clearly that the word of Exodus 20:15 has been disobeyed. That is, the ordinances are an attempt by human beings to weave the word of God into a concrete social context. Thus, more than the literal obedience to the word of God is required of the Israelites if they are to live as people under a theopolitical covenant. There are many examples in Western political speculation to document the transition from the tension between the city and the word of God into a dogmatic literalism that collapses the distinction between law and justice, the word and the ordinance, the heavenly and the earthly city, and so on. As we shall see, there are close analogues in the transition from puritanical and dogmatic Islam to Islamist terrorism.

In this chapter, however, we will indicate only the Western endpoint: the ideological obliteration of these well-understood distinctions under conditions of totalitarian domination enforced by terror. Specifically, Arendt's discussion of the "novel form of government" that combined ideology and terror provides the most complete and accessible account of the external aspects of the phenomenon. Totalitarian domination is not tyranny—a lawless, arbitrary government where power is both concentrated and exercised in the interests of a single individual.

To begin with, and notwithstanding the notorious tendency of totalitarian regimes to disregard their own laws, totalitarians claim to be executing strictly, directly, and unequivocally the laws of History, of Nature, or of God. In nontotalitarian contexts, of course, History, Nature, and God (and especially the last two) have been invoked as the sources or the ground from which positive laws, decrees, and ordinances have sprung. Likewise, totalitarian rule claims to be immediately obedient to those same laws, the defiance of which is understood to be the utmost in arbitrariness. Nor is rule exercised in anyone's interest: on the contrary, obedience to these suprahuman laws is precisely what enables totalitarians to sacrifice everyone's interest in order to execute and enforce the "higher" law directly.

Laws in the sense of legislation or positive law have as their focus the variegated behaviour of human beings. Accordingly, they are particular, pragmatic, and historically circumstantial. Standing apart from positive law or "above" it, the tradition of "divine" or "natural" law expresses a general source of authority for right and wrong that is applied in each case, but that no one case embodies.

Totalitarian "lawfulness," in contrast, attempts to translate directly the authority of right and wrong into specific cases but at the same time to retain its generality. The direct execution of the laws of History, Nature, or God does not apply to particular individuals or even to particular classes, races, or religions, but to humanity as a whole.

Arendt has summarized this new understanding of law in the following words: "the term 'law' has changed its meaning: from expressing the framework of stability within which human actions and motions can take place, it became the expression of the motion itself."

The most peculiar consequence that follows from this novel understanding of law is that the movement that the new law expresses is endless. "If," Arendt wrote, it is the law of nature to eliminate everything that is harmful and unfit to live, it would mean the end of nature itself if new categories of the harmful and unfit-to-live could not be found; if it is the law of history that in a class struggle certain classes "wither away," it would mean the end of human history itself if rudimentary new classes did not form, so that they in turn could "wither away" under the hands of totalitarian rulers.

In other words, the law of killing by which totalitarian movements seize and exercise power would remain a law of the movement even if they ever succeeded in making all of humanity subject to their rule. In the same way, the executors of the law of God are engaged in an endless search for the enemies of God who then may justly, piously, and directly be put to the sword.

As we shall see in detail below, the Islamist or "salafist" understanding of Islamic law, the Sharia, is that it is to be enforced directly, as the law of God, on all humanity. Because in fact this cannot be done, they too are on an endless treadmill of violence and war.

Terror, especially, is a means not to fight opposition but to create it, in order that it may then be righteously extinguished, thereby enabling totalitarians further to actualize the "higher" law. Terror marks out the enemies of humanity whether they are consciously opposed to the totalitarian movement or not.

That is, guilt and innocence in the ordinary sense have been eclipsed by the execution of judgments sanctioned by a "higher" source—Nature, History, God. Finally, there must be an account, a justification, for all the killing, a narrative that creates "objective" enemies whose existence and subsequent extinction keeps the murderous apparatus in motion. This account and justification is called by Arendt "ideology," and she gives this much-overused term a clear but also somewhat idiosyncratic meaning.

A STRATEGIC RESPONSE TO TERRORISM

Many observers have suggested ways that the United States might respond to the terrorist attacks of 11 September 2001. Some of these proposals have been presented in a systematic manner. Many of these suggested strategies are threat-based, and most are directed specifically at the al-Qaeda network. While any effective strategy must address existing threats, developing a strategy based on one type of event can lead one to focus on the tactical and operational aspects of terrorism, perhaps missing some of the more important strategic dimensions. A threat-based perspective often produces only a short-term response. Thus, policies based on recent experience may miss important options.

Other studies suggest conceptual approaches that are objectives-based. One widely cited and relatively comprehensive framework has been developed by the ANSER Corporation; it has been disseminated widely via its online Homeland Security journal and newsletter. It offers an objectives-based approach that provides a phased response to a range of threats.

It posits a strategic cycle of deterrence, prevention, preemption, crisis management, consequence management, attribution, and response. Another systemic approach is contained in the recently crafted United States National Security Strategy. It outlines an immediate response to terrorism: a short-term campaign to disrupt and destroy terrorist organizations through direct action, preemption, and denial of sponsorship, support, and sanctuary. In the longer term it calls for a "war of ideas" to criminalize and isolate the terrorists, coopt the support of moderates, diminish the underlying conditions of violence, and employ public diplomacy to provide information, truth, and hope to the citizens of societies from which terrorism rises. And underlying both of these policies is the effort to protect and defend the United States through homeland security initiatives.

The framework here incorporates many of these previous ideas about what constitutes an effective counter-terrorist strategy, while expanding them to respond to a more systematic conception of terrorism. The framework outlined here is time-phased from pre-attack prevention, through trans-attack mitigation to post-attack response. It also incorporates strategies to foster long-term dissuasion of terrorist attacks and assurance. It focuses on terrorism as a tactical, operational, and strategic issue and on how to defeat terrorist attacks against the United States and the appeal of terrorists to their target audience abroad. It aims to marginalize the terrorist message and defeat the terrorist's strategy.

PRE-ATTACK COMPONENTS

A systematic pre-attack, anticipatory response to terrorism might incorporate four complementary elements, each directed against an essential component of terrorism. Many of these actions constitute the immediate response to 9/11. The four suggested here are preemption, protection, and preparation, all toward prevention.

- Preemption strikes at terrorist operations: motivations to act, organizational structure and organizational support activities, and especially the terrorist infrastructure. Specific actions to blunt, divert, deflect, or arrest an attack are possible by limiting critical support, sanctuary, planning, communications, and movement. Even the flattest terrorist organizational structure with the fewest nodes of contact has a vulnerable critical support infrastructure. With targeted intelligence, overt or covert government action can block operational terrorist activities.
- Protection is physical security and other prophylactic measures that can limit the extent of the damage produced by specific terrorist weapons and tactics. Hardening and denial-of-access designs can protect high-risk facilities. When confronted with a hardened or secure target, terrorists will usually go elsewhere to seek a softer avenue for attack. Broadening the scope of effective protection to a wider set of potential targets will continue to complicate terrorists' plans. Even access control procedures, if exercised and advertised, can deflect terrorists to other targets. These efforts also can help prepare the targeted national government or citizenry against the psychological impact of terrorism.
- Preparation must be aimed at the ultimate targets of the terrorist strategy. Terrorism is a physical attack intended to produce a psychological effect; an effective counter-strategy must have a psychological component. Cases of successful preemption and prevention, as well as heightened protection measures, must be made public to send clear messages to both the terrorist and the target. Letting the terrorist know that we are prepared will act to multiply the deterrent effect of preventive and protective efforts. And increased citizen knowledge and participation will shore up confidence and resolve, and help blunt the impact should an attack occur. Public involvement in preparatory exercises and education programs will impart the protective power of knowledge to citizens and officials.

Prevention of the full range of terrorist attacks is extremely unlikely, but even one success will save lives and property, and it will contribute to the success of the overall counter-terror strategy. Identification of terrorists' plans, detection of repeated surveillance of potential objects of attack, or knowledge of weapons construction activities can lead to direct intervention to prevent specific acts, or at least categories of acts. General or targeted warnings and increased vigilance can be effective. By stopping individual acts, limiting the extent of the damage that these acts can inflict, and bounding the extent of the psychological effects produced by an attack, prevention will limit the terrorists "success" and thus the attractiveness of terrorism.

7

Pakistan's Spy Agency and Terrorism

INTRODUCTION

The murder of a prominent Pakistani journalist, Syed Saleem Shahzad, who was kidnapped last Sunday in Islamabad after repeated threats by Inter-Services Intelligence, Pakistan's premier spy agency, should be a clarion call to the international community about the increasingly strident and lawless behaviour of certain elements operating freely inside Pakistan's military and intelligence organizations. Shahzad's abduction occurred a few days after he wrote a forthright article suggesting that the militant attack on Pakistan's main naval base near Karachi on May 22 was in retaliation for an army crackdown on al Qaeda cells infiltrating its inner sanctum. His murder was clearly a warning to other seekers of truth that ISI thuggery knows no limits. (Of course, the ISI has denied any involvement in the killing.)

Questions about the ISI's role in Pakistani society have intensified in the aftermath of the U.S. raid on Osama bin Laden's Abbottabad compound, where he had been hiding in plain sight for the past six years. Further doubts about the ISI's intent and motives were raised in a Chicago courtroom last week when witness testimony linked the ISI to the Pakistani terror group, Lashkar-e-Taiba, that allegedly carried out the 2008 Mumbai terrorist attacks. One hundred and seventy people died in that three-day siege of India's largest commercial hub.

The finger of responsibility in these recent events often points to a shadowy outfit of the ISI dubbed the S-Wing. A notorious group of operatives, the S-Wing is made up of active ISI officers, recent retirees, and plain-clothes civilians with highly specialized training—all dedicated to protecting and preserving Pakistan's territorial integrity using any method, at any cost, with no regard for collateral damage. As black-ops units go, it is about as thuggish and ruthless as is possible, without being a criminal organization.

That is why the S-Wing should be declared a sponsor of terrorism under the "Foreign Governmental Organizations" designation by the U.S. State Department. It no longer matters whether the ISI is willfully blind, or explicitly complicit, in the murderous plots attributed to the S-Wing, which the ISI routinely denies any knowledge of or responsibility for. S-Wing must be stopped dead in its tracks before immeasurable harm comes from the missionary zeal of its agents, no matter how misguided their mission may be.

What has become clear in the weeks since bin Laden was shot dead on Pakistani soil, and a nation stood shocked and humiliated, is that the people of Pakistan are not America's enemies.

What has become clear in the weeks since bin Laden was shot dead on Pakistani soil, and a nation stood shocked and humiliated, is that the people of Pakistan are not America's enemies. They know less than we do about the machinations of their spymasters, and are powerless to do anything about it. Neither is Pakistan's incompetent and toothless civilian government America's enemy, even if its politicians and officials love to line their pockets with our taxpayer dollars while disingenuously shielding the ISI from public scrutiny.

The enemy is the ISI—it runs Pakistan from the shadows like a puppet master. The ISI is a danger to civilized societies everywhere, because it nurtures and breeds hatred among Pakistan's Islamist masses, and then uses their thirst for jihad as a foreign policy sledge hammer against Pakistan's neighbors and allies, often for no purpose besides just creating chaos. Its financial, logistics, and intelligence support of myriad jihadist groups in Kashmir thrice brought India and Pakistan to war—once a near-nuclear confrontation. Its S-Wing planning and logistics support for Afghanistan's Taliban from Pakistani soil has so muddied the waters in Afghan-Pakistan relations that President Hamid Karzai turned to India to counterbalance Pakistan's negative influences on his country. War, against both its neighbors, may be big business for Pakistan's army and intelligence organs, but the damage to civil life is nearly catastrophic—and it must now stop.

That the ISI even operates a special unit like the S-Wing to do the nation's dirty work demonstrates Pakistani state malfeasance in a manner that can no longer be ignored, compromised with, or tolerated by the U.S. government—indeed, by any government anywhere. What's worse, there is a growing likelihood that the ISI no longer controls the monsters it has created within its highly insulated and cellular organizational structure.

The time has come for America to take the lead in shutting off the political and financial support that gives life to an organ of the Pakistani state dedicated to undermining global anti-terror efforts. The ISI embodies the scourge of radicalism and Islamist terror that emanates from the soil it runs roughshod over. It defies common sense that a country that produced the Muslim world's first Nobel laureate in 1979, a president of the International Court of Justice at The Hague in 1970, and a rock-star musician in the 1990s, can allow the thuggery and criminality that nurtured Khalid Sheikh Mohammed's evil genius, harboured Osama bin Laden and other al Qaeda criminals, sold its nuclear secrets for profit, and murdered its brightest democracy-loving politicians, to define its identity as a nation.

We must help Pakistan by ending the ISI's reign of terror over its own citizens, and giving the ordinary people of Pakistan a chance to take their country back from the thugs who rule over them.

PAKISTAN'S TERROR TIES AT CENTER OF UPCOMING CHICAGO TRIAL

It may be years, if ever, before the world learns whether Pakistan's powerful Inter-Services Intelligence Directorate (ISI) helped hide Osama bin Laden.

But detailed allegations of ISI involvement in terrorism will soon be made public in a federal courtroom in Chicago, where prosecutors last week quietly charged a suspected ISI major with helping to plot the murders of six Americans in the 2008 Mumbai attacks.

The indictment has explosive implications because Washington and Islamabad are struggling to preserve their fragile relationship. The ISI has long been suspected of secretly aiding terrorist groups while serving as a U.S. ally in the fight against terror. The discovery that bin Laden spent years in a fortress-like compound surrounded by military facilities in Abbottabad has heightened those suspicions and reinforced the accusations that the ISI was involved in the Mumbai attacks that killed 166 people.

"It's very, very troubling," said Congressman Frank Wolf, R-Va., chairman of the House Appropriations sub-committee that oversees funding of the Justice Department. Wolf has closely followed the Mumbai case and wants an independent study group to review South Asia policy top-to-bottom.

"Keep in mind that we've given billions of dollars to the Pakistani government," he said. "In light of what's taken place with bin Laden, the whole issue raises serious problems and questions."

Three chiefs of Lashkar-i-Taiba, the Pakistani terrorist group, were also indicted in Chicago. They include Sajid Mir, a suspected Mumbai mastermind whose voice was caught on tape directing the three-day slaughter by phone from Pakistan. Mir, too, has links to the ISI. He remains at large along with the suspected ISI major and half-a-dozen other top suspects.

Despite the unprecedented terrorism charges implicating a Pakistani officer, the Justice Department and other agencies did not issue press releases, hold a news conference or make any comments when the indictment was issued last week. The 33-page documentnames the suspect only as "Major Iqbal." It does not mention the ISI, although Iqbal's affiliation to the spy agency has been detailed in U.S. and Indian case files and by anti-terror officials in interviews with ProPublica over the past year.(

"Obviously there has been a push to be low-key," said an Obama Administration official who spoke in an interview last week and requested anonymity because of the pending trial. "There is a desire to make sure the handling of the case doesn't mess up the relationship" with Pakistan.

The first public airing of the ISI's alleged involvement in the Mumbai attack will begin on May 16 with the trial of Tahawwur Rana, owner of a Chicago immigration consulting firm. Rana was arrested in 2009 and charged with material support of terrorism in the same case in which the four suspects were indicted last week. The star witness will be David Coleman Headley, a Pakistani-American businessman-turned-militant who has pleaded guilty to scouting targets in India and Denmark. Rana allegedly helped Headley use his firm as a cover for reconnaissance.

Rana's attorney, Charles Swift, contends that Rana is not a terrorist because he thought he was assisting the ISI with an espionage operation. Swift said the U.S. indictment omits the ISI in hopes of mitigating tensions.

"The U.S. is attempting to walk a fine line between disclosure and non-disclosure," Swift said. "What's unusual is that the reason is to protect diplomatic relations... This indictment answers a few questions, but like everything else in this case, it raises even more."

Even before the bin Laden slaying, the Obama Administration had taken a tougher tone about the ISI's alleged links to militants. But a U.S. official said this week that U.S. counter-terror agencies still think that any involvement in the Mumbai attacks was limited to rogue officers.

"No one is saying we can't work with the ISI—people are just pointing out the problems that exist," said the official, who requested anonymity because of the sensitivity of the issue. "I think the problems are largely with individual officers as opposed to the institution."

Pakistani officials deny that the security forces were involved in Mumbai. A senior Pakistani official questioned the credibility of Headley, who was an informant for the Drug Enforcement Administration when he began training with Lashkar in 2002.

"When somebody is a double agent, whatever he says in a U.S. court is not credible from our perspective," said the official, who requested anonymity because of the pending trial. "There is no Major Iqbal serving in the ISI who has been involved in the Mumbai attacks."

Headley has opened a door into an underworld in which spies, soldiers and terrorists converge. Although most of the prosecution's documents in the voluminous Chicago court file remain sealed, a recent judge's ruling in the Rana case says Headley admitted to working for the ISI as well as for Lashkar and al Qaeda.

"I also told [Rana] about my meetings with Major Iqbal, and told him how I had been asked to perform espionage work for ISI," Headley testified, according to the April 1 document. "I told [Defendant] about my assignment to conduct surveillance in Mumbai…I told him that Major Iqbal would be providing money to pay for the expenses."

Headley described an almost symbiotic bond between Lashkar and the ISI, which helped create the group as a proxy army against India. His account has been corroborated through other testimony, communications intercepts, the contents of his computer and records of phone and e-mail contact with ISI officers, anti-terror officials say.

Senior ISI officers served as handlers for Lashkar chiefs and provided a boat, funds and technical expertise for the Mumbai strike, according to a 119-page report by India's National Investigation Agency on its interrogation of Headley last year in Chicago.

Headley trained in Lashkar camps before being recruited in 2006 by an ISI officer, Major Samir Ali, who referred him to Iqbal in Lahore, the report says. Iqbal became Headley's handler, introducing him to a Lt. Col. Shah and giving him months of spy training before deploying him to India, according to the Indian report, which officials say repeats Headley's confessions to the FBI.

The U.S. indictment alleges that Iqbal gave the American $28,000 for the front company in Mumbai and other expenses.

Iqbal and Mir directed Headley's scouting of luxury hotels and other targets chosen to ensure that Americans and other Westerners would die. The two handlers met separately with Headley to discuss missions and receive his videos and reports, the indictment says. Iqbal took part in the decision to hit a Jewish center run by an American rabbi, who was killed along with his pregnant wife, according to the Indian report and U.S. investigators.

Headley also met at least twice with Iqbal in late 2008 to launch a Lashkar plot against a Danish newspaper that had printed cartoons of the Prophet Mohamed, according to the Indian report and investigators. Prosecutors charged Mir in the Denmark case but did not mention the suspected role of the major, who ended contact with Headley in early 2009 when Lashkar put the plot on hold, according to court documents.

But Headley stayed in touch with the other two ISI officers as he continued the Denmark plot for al Qaeda, according to investigators. His al Qaeda interlocutor was allegedly a well-connected former Pakistani Army major, Abdur-Rehman Syed, whose ISI handler was Col Shah and who had contacts with bin Laden, the report says.

"Rehman is directly in touch with the top...of al Qaida including Ilyas Kashmiri who is now the number 3 in the al Qaida hierarchy in Pakistan," the report says. "Rehman has met Osama a number of times. [Rehman] once told Headley that his set up has been given the name...Army of Fidayeens by Osama bin Laden himself."

Sajjan Gohel of the Asia-Pacific Foundation, a London security consulting firm, pointed to another possible link between the Mumbai case and bin Laden. The spy agency's director during the period that the Mumbai plot developed was Gen. Nadeem Taj. Two months before Lashkar struck Mumbai in November 2008, Taj stepped down, reportedly as the result of U.S. pressure.

Before taking leadership of the ISI, Taj was commandant of the military academy in Abbottabad, the city where bin Laden was found on Sunday. Taj has been sued in federal court in New York by families of the victims of Mumbai for his alleged role in their deaths.

Gohel said the United States and Pakistan are "moving from 'frenemies' to outright enemies." "If the ISI were involved in protecting bin Laden, that means they were capable of protecting any terrorist in Pakistan," he said. "It also means US citizens were acceptable targets. It illustrates the fact that since 9/11 the ISI has been duplicitous, disingenuous and potentially allowed acts of terrorism to be exported from its territory."

Despite the increasing tensions with Pakistan, the U.S. official credited the ISI with helping in the hunt for bin Laden. "There are lots of pieces of evidence that got us to where we are today," the official said. "Some of those pieces were facilitated by the ISI. We have to look at the full scope of our relationship."

That relationship can survive, the Pakistani official said. "Both countries are allies, and important allies," he said. "I think our relationship is beyond Headley's statements in court."

MAJORITY APPROACH TO THE THREAT OF TERRORISM

Lord Bingham's lead judgment represents the *ratio decidendi* as it had the agreement of six of the Lords. Unlike Lord Hoffmann, Lord Bingham was not prepared to hold that no public emergency threatening the life of the nation existed. Nevertheless, he upheld the appeal on the grounds that the detention powers were disproportionate and discriminatory. In relation to the *designation* issue, Lord Bingham's approach essentially absolved the Government from advancing clear and convincing evidence to Parliament (and the courts) to demonstrate that a public emergency threatening the life of the nation actually existed. Lord Bingham approved and applied the case law of the ECrtHR on Article 15 ECHR granting a wide margin of appreciation. He found that to hold that there was no public emergency in cases where, 'a response beyond that provided by the ordinary course of law was required, would have been perverse'.

This reasoning, however, is illogical as it essentially bases the determination of the question of whether a public emergency exists on the measures taken to address it. As Tom Hickman has observed, 'if one is to infer from the fact that exceptional measures have been taken that such measures are legitimate then the criteria of legitimacy (i.e., public emergency) is relieved of substance'. Lord Bingham went on to hold that it was for the appellants to demonstrate that the Government's claim that there was an emergency that required derogation from the ECHR was 'wrong and unreasonable'. The appellants, however, had 'shown no ground strong enough to warrant displacing the Secretary of State's decision on this important threshold question'. Lord Bingham's reasoning is highly problematic. This reversal of the burden of proof in relation to the existence of a public emergency threatening the life of the nation raises serious concerns from a purely practical perspective. It is difficult to see how individuals will ever be able to disprove the government's view that an emergency exists, not least because the relevant evidence will be in the hands of the government. Lord Bingham's view also runs contrary to the approach taken by the ECrtHR. As indicated earlier, the Strasbourg authorities have repeatedly confirmed that the burden is not upon the individual, but upon the government to demonstrate that there exists a national emergency that requires derogation from international human rights obligations. It is noteworthy that the Human Rights Committee in its *General Comment 29* has taken a similar view.

With regard to the *interference* issue, Lord Bingham held that the detention power was not rationally connected to the objective of addressing the imminent threat of terrorism as it did not correspond to that objective in several respects.

First, and assuming that the terrorist threat constituting a national emergency stemmed from Al Qa'ida, the detention power set forth in s 23 ATCSA powers applied to non-Al Qa'ida terrorists as well.

Second, it applied to Al Qa'ida supporters who posed no direct threat to the national security of the UK.

Third, it did not apply to the threat from terrorists who were UK nationals.

And fourth, it allowed any 'suspected international terrorist to leave our shores and depart to another country, perhaps a country as close as France, there to pursue his criminal designs'. This, said Lord Bingham, was 'hard to reconcile with a belief in [the terrorists'] capacity to inflict serious injury to the people and interests of this country'. As a result, the measure taken (i.e., s 23 ATCSA) was not strictly required by the exigencies of the situation and the derogation was hence unlawful.

From a purely logical perspective, Lord Bingham's reasoning with regard to the *interference* issue is not entirely consistent, especially in light of his findings in relation to the *designation* issue. He essentially held that it was not for the Court but for government to assess whether the threat of terrorism constituted a public emergency.

Nonetheless, Lord Bingham then went on to hold that s 23 ATCSA was not rationally connected to the emergency and thus not suitable to reduce the imminent threat. He failed to explain, however, how he was able to conclude that a measure was not connected to the national emergency, or not suitable to reduce the imminent threat, when the nature and quality of the threat itself was not something that the Court was able to examine or determine. This is a logical gap in the majority decision of the House of Lords in *Belmarsh Detainees* and also in the case law of the ECrtHR in the area of emergency derogations more generally. From a logical standpoint, it is simply impossible to determine whether an emergency measure is suitable to address a threat or a crisis without establishing what the nature or quality of the threat or crisis is in the first place.

THE DISSENTIENT APPROACH TO THE THREAT OF TERRORISM

Lord Hoffmann, on the other hand, chose to undertake an examination of the quality and nature of the threat of terrorism to the UK and found that it did not constitute a 'war or other public emergency threatening the life of the nation'. He further held that it was insufficient merely to produce evidence of a credible plot to commit terrorist outrages since that did not meet the need to show that the threat of terrorism constituted a public emergency threatening the life of the nation. According to Lord Hoffmann: The Armada threatened to destroy the life of the nation, not by loss of life in battle, but by subjecting English institutions to the rule of Spain and the Inquisition. The same was true of the threat posed to the United Kingdom by Nazi Germany in the Second World War. This country, more than any other in the world, has an unbroken history of living for centuries under institutions and in accordance with values which show a recognisable continuity.

The events of 11 September 2001 in New York and Washington and 11 March 2003 in Madrid make it entirely likely that the threat of similar atrocities in the United Kingdom is a real one ...This is a nation which has been tested in adversity, which has survived physical destruction and catastrophic loss of life. I do not underestimate the ability of fanatical groups of terrorists to kill and destroy, but they do not threaten the life of the nation. Whether we would survive Hitler hung in the balance, but there is no doubt that we shall survive Al Qaeda. The Spanish people have not said that what happened in Madrid, hideous crime as it was, threatened the life of their nation. Their legendary pride would not allow it. Terrorist violence, serious as it is, does not threaten our institutions of government or our existence as a civil community.

Lord Hoffmann explicitly held that there were legal limits to the Government's capacity to determine when a situation of public emergency existed, and further, that the Government was in fact wrong to declare a situation of public emergency in the aftermath of 9/11. In doing so, he did not grant the Government a wide margin of appreciation with regard to the *designation* issue.

As a consequence, he also did not address the question of whether the Government's measures adopted in s 23 ATCSA were 'strictly required' (ie, the *interference* issue). What is remarkable about Lord Hoffmann's judgment is that he is able to determine, without access to specific intelligence information, that the current threat of terrorism to the UK does not threaten the life of the nation.

In fact, he explicitly accepts that there is a serious terrorist threat to the UK. But this threat is put into perspective by drawing comparisons both to historical threats to the UK and more recent manifestations of terrorism like the 9/11 attacks and the Madrid train bombings. And so the Government's general policy decision about the nature and quality of the threat of terrorism is submitted to judicial scrutiny despite lack of access to specific intelligence information.

PRINCIPLE IN THE CONTEXT OF ANTI TERRORISM LAWS

A key question in the political and academic discourse on the legislative response to the threat of international terrorism has been the question of proportionality. While some have argued that the laws enacted to counter terrorism strike the right balance between national security imperatives and concerns for civil liberties and human rights, others have regarded them as disproportionate and as an overreaction. What both sides have in common, however, is that they generally approach the question of proportionality without examining the nature and quality of the terrorist threat and by accepting the executive's assertion that the threat may warrant a range of comprehensive counter-measures.

What proportionality generally requires is that there is a reasonable relationship between the means employed and the aims sought to be achieved. Essentially proportionality requires one to determine whether a measure of interference, which is aimed at promoting a legitimate public policy, is either unacceptably broad in its application or has imposed an excessive or unreasonable burden on certain individuals. A decision that takes into account proportionality principles should, inter alia, impair the right in question as little as possible, be carefully designed to meet the objectives in question, and not be arbitrary, unfair or based on irrational considerations. In order to establish whether counter-terrorism laws and measures meet the objectives in question it is imperative to identify clearly what those objectives are. The objective of anti-terrorism laws is, in most cases, the reduction of the threat of terrorist attacks or activities. Thus it is logically necessary for a thorough proportionality analysis to consider or assess the quality and nature of the threat. I would argue that in the absence of such analysis, any proportionality assessment is incomplete.

Nonetheless, both the European Court of Human Rights (ECrtHR) and national courts, most recently the House of Lords, have taken a deferential approach and granted national authorities a wide 'discretionary area of judgment', or, in the terminology of the ECrtHR, a 'wide margin of appreciation' with regard to the existence and analysis of the threat of terrorism that may constitute a so-called 'public emergency'.

One rationale behind this deferential approach, especially in common law countries, seems to be that in terms of both constitutional competence and expertise in the area of national security it is for government (and perhaps Parliament) rather than the courts to assess whether a public emergency exists.

While not addressing the constitutional implications of this position, the context of international terrorism this rationale is flawed in its logic.

Courts can and should be in a position to assess the nature and size of the terrorist threat without necessarily having to have access to specific intelligence. This is not to say that courts should not have access to specific intelligence or classified information held by the government. On the contrary, access to such information may be essential to fulfil fair trial requirements in proceedings against persons accused of terrorism offences. However, the difficulties and challenges that classified information poses for the courts shall not be the subject of analysis here. This chapter is that in spite of any access to specific intelligence information, courts can and should submit general policy decisions about the threat of terrorism to judicial scrutiny.

The argument has both an international and a domestic dimension (although the domestic dimension is related to the international one).

First we will argue that developments in international human rights law provide ample justification for an 'extension' of the competency of the courts — especially the ECrtHR — to assess the nature and quality of the terrorist threat that is seen to constitute a 'public emergency'. Second, we will argue that an 'extension' of competency of domestic and national courts would also be possible and desirable and, further, that it would also be the logical consequence of findings by the House of Lords in the *Belmarsh Detainees* decision of December 2004.

WINDS OF HATRED CONTINUE TO SWEEP ACROSS PAKISTAN

At a time when winds of change have been sweeping across many Islamic countries with calls for greater freedom and democracy, winds of hatred continue to sweep across Pakistan.

Pakistan, which has become over the years a breeding ground of Islamic beliefs of the most irrational and extreme kind, is helpless in the face of these winds of hatred. These winds have distorted Islam beyond recognition and provided a breeding ground for Islamic extremism and jihadi terrorism of various hues. More murders and more crimes of various kinds are committed in Pakistan in the name of and for the sake of Islam than in any other Islamic country of the world. Unless and until the breeding grounds of hatred from where these winds rise are eliminated, extremism and terrorism will continue to find nourishment in the soil of Pakistan. No amount of change in the other Islamic countries through which the winds of change have been sweeping would provide relief to the rest of the world from the scourge of Islamic extremism and jihadi terrorism. It is a plague over which the state of Pakistan has no control.

This plague claimed one more fatal victim on March 2,2011, when Shabaz Bhatti, a Christian belonging to President Asif Ali Zardari's Pakistan People's Party (PPP), who was holding the Minorities Affairs portfolio, was shot dead by a group of unidentified assassins as he was being driven to work in Islamabad. Reports from Islamabad indicate that he was travelling in his official car without being escorted by a security team despite the fact that he was one of the most threatened members of the Council of Ministers because of his criticism of the blasphemy law which provides for a mandatory death penalty to anyone insulting the Holy Prophet.

The law has come in for strong criticism from liberal elements in the rest of the world because the rules of evidence governing trials under it are so flimsy that anyone can accuse anyone of insulting the Holy Prophet and get that person convicted without satisfactory corroborative evidence. Previously, only insulting the

Holy Prophet was considered an act of blasphemy, now, even criticising the law is treated an act of blasphemy by extremist elements which do not hesitate to kill anyone criticising the law. This was the second high-profile assassination this year of persons criticising the law.

In January, Salman Taseer, a liberal Muslim, who was the Governor of Punjab, was assassinated by one of his own security guards because he dared to criticise the law and visited in jail a Christian who had been convicted under the law. Death threats have reportedly been held out against Mrs.Sherry Rehman, a Member of Parliament belonging to the PPP, for allegedly suggesting a re-look at the law. How can one save Pakistan from the clutches of Islam of the most extreme kind when the assassin of Taseer was not condemned as a murderer, but was hailed as a saviour of Islam by some sections of the population, including lawyers? Organisations such as the Tehrik-e-Taliban Pakistan (TTP), which has claimed responsibility for the assassination of the Christian Minister, and the Lashkar-e-Jhangvi (LEJ), the extremist Sunni organisation, look upon it as a God-ordained duty to eliminate anyone who is seen by them as insulting Islam even for criticising the obnoxious features of it in Pakistan.

One cannot hope for any salvation for Pakistan from these winds of hatred unless there is a mass uprising to break the stranglehold of these elements over the society and the State. Pakistan is a State governed by fear—not the fear of despots, but the fear of the irrational clergy and even more irrational extremist organisations. Unless the people are able to rid themselves of this fear and come out in the streets against these organisations, the winds of hatred will continue to blow across the country.

Unfortunately, Pakistan is a country where liberalism is merely a talking point in the drawing rooms of the elite and not a rallying cry for protests in the streets against the irrational and extremist elements. It is a gloomy situation from which no exit is in sight.

Politics of Feudalism

The course of politics since independence has been determined and dominated by a small segment of society and nothing has happened during the half century of economic turbulence and social chaos to alter the class composition of the leadership, which still comes from the feudal-army-bureaucracy conglomerate. Feudal system in Pakistan continues to exist and flourish with all its evils and colonial legacies which have virtually disfranchised the bulk of the population by its monopoly of power. This system creates areas of oppressive influence for the feudal lords, particularly in the vast rural areas constituting 70 per cent of country's population. With only two per cent of the population the feudal lords are able to capture bulk of the assembly seats, thus denying the poor and middle class their legitimate share in the government.

From 1947 to 1970, 75 per cent of the members of our parliaments were feudals and the remaining 25 per cent who owed their parliamentary carrier to their place and position in the services and professions, had also come from a feudal background. The feudal families which had been discarded by the electorate in the impartial elections held in 1970 by General Yahya Khan returned to the assemblies in the non-party basis elections held by General Zia in 1985. Landowners captured 117 seats out of 219 contested seats while 42 went to businessmen, Long military rule and the prolonged suspension of political processes has enabled the Pakistani ruling elite to refine the art of social engineering. The feudal lords in collaboration with the military-civilian bureaucratic system have been able to determine whom to include and exclude from the development process.

Although our feudal lords are functioning within the framework of a democracy but the republican form of government had been republican in a very narrow and restricted sense. That form persists to this day even after we have had a series of elections, following the fall of Ayub Khan and the end of the Zia regime. Soldiers by profession, they were nonetheless landowners by heredity or acquisition. The feudal lords have reinforced their power manifold by dominating industry, business and politics. Once the stronghold of the feudals, the national and provincial assemblies are now the property of both the landlords and the capitalists. Worse, the system has been consolidated by the induction of retired personnel of the armed forces and civil service into its ranks. Thus, feudal individuals, higher-echelon civil servants and the military establishment are tied into a single web. The system excludes the mass of the people from the concerns of the state and depends upon the loyalties and manipulations of a relatively small number of families and groupings. It breeds narrow relationships and discretionary favors, rather open and transparent system.

Genuine and effective land reforms have always been resisted by the feudal lords irrespective of the democratic or authoritarian nature of the regime. In times of army rule, feudals join the ruling junta to protect their system from being discontinued. And the junta acquiesces because it badly needs the feudal clout to keep the teeming millions in the rural areas from revolting against their dictatorial practices. And in times of democratic dispensation, the fat cats of the system, using their feudal hold on these very teeming millions return to the legislative assemblies where they to it that no law is made to loosen the hold of the feudal system on the national economy.

The elected representatives on either side of the aisle would, while quarreling with great passion and violence on non-issues, remain united on preserving the status quo. Sworn enemies share common ground against reform of land, labour, and taxation; against liberation of laws and the independence of judiciary; against devolution of power and distributive justice. The country's constitution was dangerously deformed by seven amendments, then an eighth one. One does not hear any more of abrogating even this last constitutional atrocity. Worst of all, the feudals would not allow even a census, for it would certainly change the political balance. The last census was held in 1981. Nominally, we have a parliamentary system of government. In fact, because of the underlying power structure, the system is anything but democratic. It takes a few million rupees to contest a seat in the National Assembly, votes are bought and sold. For all practical purposes, the people are largely disenfranchised. Parliamentary elections are a game of musical chairs among the rural and urban elite and their hangers-on. For the common man and Haris living in the villages, therefore, it hardly amounts to any visible change in the operating conditions around them or any material difference in the quality of their life after the creation of Pakistan. The gap between the propertied and propertiless classes instead of narrowing, has widened still further

CURRENT ISLAMIC SUICIDE TERRORISM

The decision to switch to suicide operations is not arbitrarily. It is important to ask two questions both of them related to the individual Islamic terrorist organisation; when do they start and when do they stop?

Egyptian Islamic Terrorist Organziations

In the mid-1990s two Egyptian Islamic terrorist organizations conducted a few suicide operations and then stopped. Ga'maa Islamyya concluded that although this method provided a tactical advantage it would alienate the group from its local support base in Egypt where this extreme measure would not be acceptable. The EIJ took a different view and realized that this was to be a very efficient tool in a global Jihad that was to be formulated together with the Al Qaeda leadership. Interestingly, the Ga'maa Islamiyya has as recently as 2003 disavowed suicide terrorism as un-Islamic and claimed that it counter productive in the furtherance of any Islamic agenda.

The Shi'a Campaign

The realization by Islamic terrorist organizations that a spectacular suicide operation is a strategic equalizer has occurred in several phases. Suicide terrorism began as a symbolic act of resistance and defiance by the Shi'as of Lebanon to enable them to fight superior enemies.

The concept of martyrdom on the battlefield was part of the cultural and religious heritage of the Lebanese Shi'as, but it took an outsider force to manipulate and invigorate these sentiments into violent action. Both Iran and their Lebanese proxies were surprised at the effectiveness of a few suicide attacks. The Shi'a terrorist organizations ceased to field suicide bombers after Israel withdrew from Lebanon in 2000. This was the result of a decision by the leaders of the organizations who acknowledged that the suicide operations had indeed worked to defeat the enemy, but also that is was a very problematic issue to legitimise from a religious perspective.

The Al Qaeda Campaign

The Al Qaeda essentially reached the same conclusions as the Hezbollah, but employed an entirely different approach in its strategic perspective.

Over the years suicide operations has become the preferred method of operation for Al Qaeda, and this is the result of the realization that it serves its purpose. Realizing that the network of Islamic terrorist organizations could not change the world on their own, Al Qaeda perceived itself as the vanguard of a global transformation that would lead to a final confrontation between righteous Muslims and everyone else.

Because Al Qaeda has limitations, it opted for carrying out high-profile terrorist operations to get the attention of all Muslims, and hope for their support in a general uprising. The developments in international affairs after September11, 2001, has proved the validity of this argument from the perspective of Al Qaeda that sees a worldwide conflict as inevitable and imminent.

Several Benefits

There are several benefits to the terrorist organization that decide to use suicide operations. (IDF 2002), (Sprinzak 2000). From an operational perspective there is the obvious advantage of not needing an escape plan. The survival of the perpetrator is considered a failure.

Other operational tactics carry the inherent risk of capture of the operators by the authorities. A captured terrorist may reveal the identity of other cell members and other information considered vital for the survival of the terrorist organization. Another operational benefit is the accuracy in bomb placement and the flexibility in the timing of the detonation. Virtually no other method of delivery ensures the correct placement of an explosive device in comparison with the adaptability of a dedicated individual. If the chosen target does not have the desired density of civilians, it is possible for the operator simply to move to a better target. Mass casualties are the norm and extensive damage to structures is virtually guaranteed if proper target surveillance has been undertaken.

There is no need for remote control detonators that are prone to faults or detection, as the operator becomes the timing device and detonator. As evidence has shown, it is almost impossible to employ timely and accurate counter-measures to avert a suicide attack. The bomb is usually well hidden and very rarely does the operator display any visible signs of anxiety. A suicide operation is always a spectacular event and wide media coverage is guaranteed. The terrorist organization will get the attention it needs for the cause. The media coverage of the attack also serves to convey an image of extreme discipline, dedication and skill of the terrorists, thus installing fear in the public (Ganor 2002).

INDIVIDUAL SUICIDE TERRORIST AND STATE SPONSORSHIP

The individual suicide terrorist is controlled by his own terrorist organization. One level above the terrorist organizations are the countries that support, encourage or legitimise the use of violent action, including suicide operations. For this reason it would be incorrect to look at the terrorist organizations as independent or rogue outlaws. They regularly consult with their benefactors who are quite aware of what is being planned but decides not to intervene for political reasons. These countries think of the terrorist groups as expedient tools who are willing to carry out the dirty work to pressure other governments without having to expose themselves to possible retaliation. State sponsorship plays a very important role in regulating the behaviour of terrorist organizations. Any future predictions on the possible development of Islamic suicide terrorism must take the role of sponsoring states seriously. An understanding of how these states perceive the legitimacy and expediency of suicide operations would likely indicate how they could be pressured or convinced to abandon their support.

There is no rule without an exception, and the exception in mind is of course Al Qaeda. At various times and in varying degrees the foundation of what was to become the Al Qaeda network was sponsored and supported by the U.S., Saudi Arabia, Pakistan and Afghanistan during the Taliban regime. All of these major players, except the Taliban, have come to regret their involvement with an organization whose ideology evolves around their own destruction. The current situation is unprecedented because there is no state that can effectively influence the behaviour of Al Qaeda.

Target Selection of Suicide Terrorism

It is worth noticing the different paths of growth taken by the various terrorist organizations that have employed suicide operations with respect to target selection. During the past two decades a general trend can be observed from the selection of military and diplomatic targets towards the indiscriminate bombings of civilians. In a chronological perspective there has been a clear trend towards an emphasis on civilian and symbolic targets.

The Lebanese civil war saw atrocities in abundance, but there was a remarkable self-restraint on behalf of the Islamic terrorist organizations that resorted to suicide operations. The controversial nature of a martyrdom operation ruled out attacks on civilians. This behaviour could not be justified of legitimised in a religious context. There was of course civilians casualties, but they were never the intended target. They just happened to be in the wrong place at the wrong time.

During the early and mid-1990s a slow development took place in which there was a widening of target classes. Palestinian suicide terrorist began to target civilians on public buses and in shopping areas, while at the same time the Egyptian groups concentrate of targets associated with the Egyptian government.

This period changed dramatically with Al Qaeda's bombing of American embassies in East Africa in 1998. From then on, both the intensity and the diffusion of targets have escalated dramatically.

However, this trend is only a general phenomenon and should be scrutinized with the developments within each terrorist organization. Target selection is not necessarily a stable factor, but might be highly dynamic.

For this purpose it is useful to employ the concept of "the displacement effect" analysed in a terrorist context by Martha Crenshaw (Crenshaw 2002). The displacement effect is an expression that relates to the gradual erosion of what constitutes an acceptable target for the terrorist organization. The widened scope of the target classes may be the outcome of a slackening of moral inhibitions. What may have started as a guerrilla style campaign targeting military installations or foreign embassies could degenerate into an indiscriminate spate of bloodletting. This development can be very difficult to relate to the original motives and purposes of the terrorist organization.

8

Pakistan and State-sponsored Terrorism

INTRODUCTION

Pakistani President Asif Zardari admitted at a conference in Islamabad that Pakistan had, in the past created terrorist groups as a tool for its geostrategic agenda. Pakistan had long been accused by neighbour India, and western nations like the United States, and the United Kingdom of its involvement in terrorist activities in India and Afghanistan. Pakistan's tribal region along the border of Afghanistan has been claimed to be a "haven for terrorists" by western media and the United States Defense Secretary. According to an analysis published by the Saban Center for Middle East Policy at Brookings Institution in 2008, Pakistan was, "with the possible exception of Iran, perhaps the world's most active sponsor of terrorist groups... aiding groups that pose a direct threat to the United States." Daniel Byman, an author, also writes that, "Pakistan is probably today's most active sponsor of terrorism".

The Pakistani intelligence agency, the Inter-Services Intelligence(ISI), is believed to be aiding these organisations in eradicating perceived enemies or those opposed to their cause, including India, Russia, China, Israel, the United States of America, the United Kingdom and other members of NATO. Satellite imagery from the FBI suggest the existence of several terrorist camps in Pakistan, with at least one militant admitting to being trained in the country as part of the going Kashmir Dispute, Pakistan is alleged to be supporting separatist militias Many nonpartisan sources believe that officials within Pakistan's military and the Inter-Services Intelligence (ISI) sympathise with and aid Islamic terrorists, saying that the "ISI has provided covert but well-documented support to terrorist groups active in Kashmir, including the al-Qaeda affiliate Jaish-e-Mohammed".

Pakistan denied involvement in militant activities in Kashmir, though President Asif Ali Zardari admitted in July 2010 that militants had been "deliberately created and nurtured" by past governments "as a policy to achieve some short-term tactical objectives" stating that they were "heroes" until 9/11.

In October 2010, former Pakistan President and former head of the Pakistan Army, Pervez Musharraf revealed that Pakistani armed forces trained militant groups to fight Indian forces in Kashmir. Many Kashmiri militant groups designated as terrorist organisations by the US still maintain their headquarters in Pakistan-administered Kashmir. This is cited by the Indian government as further proof that Pakistan supports terrorism. Many of the terrorist organisations are banned by the UN, but continue to operate under different names. Even the normally reticent United Nations Organization (UNO) has also publicly increased pressure on Pakistan on its inability to control its Afghanistan border and not restricting the activities of Taliban leaders who have been declared by the UN as terrorists. Both the federal and state governments in India continue to accuse Pakistan of helping several banned terrorist organisations, including the Indian organisations unhappy with their own Government, like the ULFA in Assam.

BACKGROUND

Until Pakistan became a key ally in the War on Terrorism, the US Secretary of State included Pakistan on the 1993 list of countries which repeatedly provide support for acts of international terrorism. In fact, many consider that Pakistan has been playing both sides in the fight against terror, on the one hand, pretending to help curtail terrorist activities while on the other, stoking it.

Even the noted Pakistani journalist Ahmed Rashid has accused Pakistan's ISI of providing help to the Taliban, a statement echoed by many, including author Ted Galen Carpenter, who states that Pakistan has "assisted rebel forces in Kashmir even though those groups have committed terrorist acts against civilians".

Allegations of State-sponsored Terrorism

Author Gordon Thomas states that whilst aiding in the capture of Al Qaeda members, Pakistan "still sponsored terrorist groups in the disputed state of Kashmir, funding, training and arming them in their war of attrition against India". Journalist Stephen Schwartz notes that several terrorist and criminal groups are "backed by senior officers in the Pakistaniarmy, the country's ISI intelligence establishment and other armed bodies of the state".

According to the author Daniel Byman, "Pakistan is probably today's most active sponsor of terrorism." writing in an article published by The Australian stated, "following the terror massacres in Mumbai, Pakistan may now be the single biggest state sponsor of terrorism, beyond even Iran, yet it has never been listed by the US State Department as a state sponsor of terrorism".

Former Pakistan Ruler Pervez Musharraf has conceded that his forces trained militant groups to fight India in Indian-administered Kashmir. He confessed that the government 3 turned a blind eye because it wanted to force India to enter negotiations besides raising the issue internationally.

India has been consistent in alleging that Pakistan was involved in training and arming underground militant groups to fight Indian forces in Kashmir.

Inter-Services Intelligence and Terrorism

The ISI, has often been accused of playing a role in major terrorist attacks across the world including

- terrorism in Kashmir,
- the July 2006 Mumbai Train Bombings,
- the 2001 Indian Parliament attack,
- the 2006 Varanasi bombings,

- the August 2007 Hyderabad bombings and
- the November 2008 Mumbai attacks.

The ISI is also accused of supporting Taliban forces and recruiting and training mujahideen to fight in Afghanistan and Kashmir. Based on communication intercepts, US intelligence agencies concluded Pakistan's ISI was behind the attack on the Indian embassy in Kabul on 7 July 2008, a charge that the governments of India and Afghanistan had laid previously.

The Afghan President Hamid Karzai, who has regularly reiterated allegations that militants operating training camps in Pakistan have used it as a launch platform to attack targets in Afghanistan, urged Western military allies to target extremist hideouts in neighbouring Pakistan. In response to the millants from Afghanistan hiding in the mountainous tribal region of Pakistan. The US and Pakistan agreed to allow US Drone Strikes in Pakistan.

Several detainees at the Guantanamo Bay facility told US interrogators that they were aided by the ISI for attacks in the disputed Kashmir Region.

Links to Terrorist Groups

Pakistan is said to be a haven for terrorist groups like Al-Qaeda, Lashkar-e-Omar, Lashkar-e-Toiba, Jaish-e-Mohammed and Sipah-e-Sahaba. Pakistan is accused of giving aid to the Taliban, "which include[s] soliciting funding for the Taliban, bankrolling Taliban operations, providing diplomatic support as the Taliban's virtual emissaries abroad, arranging training for Taliban fighters, recruiting skilled and unskilled manpower to serve in Taliban armies, planning and directing offensives, providing and facilitating shipments of ammunition and fuel, and on several occasions apparently directly providing combat support," as stated by the Human Rights Watch. In 2008, the US has stated that the next attack on the US could originate in Pakistan. In June 2009, India's army chief, General Deepak Kapoor, used a meeting with US national security adviser Jim Jones to claim that Pakistan was home to 43 "terrorist camps", while rejecting suggestions of engaging in fresh peace talks. Another militant outfit, the JKLF, has openly admitted that more than 3,000 militants from various nationalities were still being trained. Other resources also concur, stating that Pakistan's military and ISI both include personnel who sympathise with and help Islamic militants, adding that "ISI has provided covert but well-documented support to terrorist groups active in Kashmir, including the Jaish-e-Mohammed." Pakistan has denied any involvement in the terrorist activities in Kashmir, arguing that it only provides political and moral support to the so-called 'secessionist' groups. Many Kashmiri groups also maintain their headquarters in Pakistan-administered Kashmir, which is cited as further proof by the Indian Government. The normally reticent United Nations Organization (UNO) has also publicly increased pressure on Pakistan on its inability to control its Afghanistan border and not restricting the activities of Taliban leaders who have been declared by the UN as terrorists.

ALLEGED PAKISTANI ARMY SUPPORT OF TERRORISTS

The United States had direct evidence that the ISI chief, Lt. Gen. Ahmed Shuja Pasha, knew of Bin Laden's presence in Abbottabad, Pakistan.

Pakistan was also responsible for the evacuation of about 5000 of the top leadership of the Taliban and Al-Qaeda who were encircled by Nato forces in the 2001 Invasion of Afghanistan. This event known as the Kunduz airlift, which is also popularly called the "Airlift of Evil", involved several Pakistani Air Force transport planes flying multiple sorties over a number of days.

According to Pervez Hoodhboy, "Bin Laden was the 'Golden Goose' that the army had kept under its watch but which, to its chagrin, has now been stolen from under its nose. Until then, the thinking had been to trade in the Goose at the right time for the right price, either in the form of dollars or political concessions".

According to a 2001 article titled "Overview of State-Sponsored Terrorism" issued by the US Office of the Coordinator for Counterterrorism, "In South Asia, the United States has been increasingly concerned about reports of Pakistani support to terrorist groups and elements active in Kashmir, as well as Pakistani support, especially military support, to the Taliban, which continues to harbor terrorist groups, including al-Qaida, the Egyptian Islamic Jihad, al-Gama'a al-Islamiyya, and the Islamic Movement of Uzbekistan."

In 2011, American troops reportedly recovered Pakistani military supplies from Taliban insurgents in Afghanistan.

Problems

US National Security Advisor James L Jones sent a message in the past to Pakistan saying that double standards on terrorism were not acceptable. In July 2010, British Prime Minister David Cameron accused the Pakistani government of double standards: "We cannot tolerate in any sense the idea that this country is allowed to look both ways and is able, in any way, to promote the export of terror, whether to India or whether to Afghanistan or anywhere else in the world."

However, UK Foreign Secretary William Hague, who was travelling with the prime minister Cameron, clarified Cameron's remarks: "He wasn't accusing anybody of double dealing. He was also saying that Pakistan's made great progress in tackling terrorism. Of course there have been many terrorism outrages in Pakistan itself."

Cameron's remarks sparked a diplomatic row with Pakistan, where he came under attack by officials and politicians who strongly criticised his comments. In December 2010, he attempted to visit Pakistan while on a tour to Afghanistan in an effort to mend relations. However, his visit was refused by Pakistan, notably as a snub to his remarks.

Afghanistan

US intelligence officials claim that Pakistan's ISI sponsored the 2008 Indian embassy bombing in Kabul. They say that the ISI officers who aided the attack were not renegades, indicating that their actions might have been authorised by superiors. The attack was carried out by Jalaluddin Haqqani, who runs a network that Western intelligence services say is responsible for a campaign of violence throughout Afghanistan, including the Indian Embassy bombing and the 2008 Kabul Serena Hotel attack.

India

The government of Pakistan has been accused of aiding terrorist organisations operating on their soil who have attacked neighbouring India. Pakistan denies all allegations, stating that these acts are committed by non-state actors.

India alleged that the recent 2008 Mumbai attacks originated in Pakistan, and that the attackers were in touch with a Pakistani colonel and other handlers in Pakistan. This led to a UN ban on one such organisation, the Jama'at-ud-Da'wah, which the Pakistani government is yet to enforce.

On 5 April 2006, the Indian police arrested six Islamic militants, including a cleric who helped plan bomb blasts in Varanasi. The cleric is believed to be a commander of a banned Bangladeshi Islamic militant group, Harkat-ul-Jihad al-Islami, and is linked to the ISI.

AL QAEDA LEADERS KILLED OR CAPTURED IN PAKISTAN

Critics have accused Pakistan's military and security establishment of protecting bin Laden, until he was found and killed by US forces. This issue is expected to worsen US ties with Pakistan.

Bin Laden was killed in what most feel was his residence for at least three years, in Abbottabad, in Pakistan. It was an expensive compound, less than 100 kilometres' drive from the capital, Islamabad, probably built specifically for Bin Laden. The compound is 0.8 miles (1.3 km) southwest of the Pakistan Military Academy (PMA), a prominent military academy that has been compared to Sandhurst in Britain and West Point in the United States. Pakistan's President Zardari has denied that his country's security forces may have sheltered Osama bin Laden.

In response to America's exposure of bin Laden's hiding place, Pakistan moved to shut down the informant network that lead the Americans there.

In addition Khalid Sheikh Mohammed, Ramzi bin al-Shibh, Abu Zubaydah, Abu Laith al Libi and Sheikh Said Masri have all been captured or killed inside Pakistan.

THE GROWING RELIGIOUS FUNDAMENTALISM IN PAKISTAN

The religious fundamentalists in Pakistan are once again on the full swing. They are very much encouraged by the outcome of the general elections in Pakistan. In overall, they have got over 16% of the total votes caste in the elections. But in the national assembly, they have emerged the second largest parliamentary group leaving behind Pakistan Peoples Party. Thanks to the support of the conservative Muslim League Nawaz MLN that their nominee for the prime minister Fazal Rehman got 86 votes while PPP trailed behind with 71 votes. The pro military Muslim League Q group was able to fetch the needed 172 votes for the Prime Minister lot.

The upsurge in the fundamentalist support was witnessed on 19th November 2002 when over 70,000 turned up for the funeral of Aimal Kansi in Quetta Baluchistan. This was the largest ever funeral in the whole history of Baluchistan. Aimal, body was brought from US where was awarded the death sentence after found guilty of killing two CIA officers in 1993. Till his last breath, he was very much proud of what he has done to the American imperialism. He has become another hero of the fundamentalists after Assama Bin Laden.

Another sign of the growth in religious fundamentalism was seen on 24th November when activists of Lashkar Tayaba, the banned religious group in Pakistan took over for six hours two temples in Jammu district of Indian held Kashmir. The incident resulted killing of the two and 10 more in the struggle to re-hold the temples by the Indian forces. There have been more incidents like this during the week in Kashmir resulting over 20 deaths mainly of Indian army men. This is no incident. The leader of Lashkar Tayaba Hafiz Saeed was released last week in Pakistan. The Pakistan authorities would not accept at Lahore High Court only a month before that he is held up with them. ? Withdraw your writ petition if you want your husband to be released? Was the message of the intelligence agencies to the wife. Later the wife of Hafiz Saeed was forced to withdraw her write petition. Hafiz Saeed as brought to his hometown after being held for over a year and was house arrested. Lahore High Court declared the move as illegal. Since his release on 18th November, he has addressed many thousands in several cities of Punjab. At Faisalabad, the third largest city of Pakistan, he spoke to over 50,000 who came to greet him after his release. Many women threw their children in the air and begged him to take them for Jihad. ? Jihad can not be stopped by any mean,? he declared at the meeting.

Lashkar Tayaba was the main religious group promoted by the military intelligence of Pakistan during the nineties to carry on jihad in the Indian held Kashmir. Many youth were provided the military training at camps set up by the group in different parts of Pakistani held Kashmir. "Suicidal missions are the only way forward to further the message of Jihad" was the main philosophy of the group training. The 11th September 2001 changed the priorities of the military intelligence. The Lashkar Tayaba was asked to wind up the camps. The leader was kidnapped. The military intelligences thought that it would be all over by these actions. That is been proved all wrong. The ghost of religious fundamentalist propped up by the military intelligence is out of control and is in a process of changing its colour. From pro military group, it is fast becoming an anti military group in Pakistan.

In the eighties, the religious fanatics were very closely linked with American imperialism. In the nineties, they left them on their own. From the beginning this century, the fanatics have taken roots among the general consciousness of the masses in Pakistan as an anti imperialist force. ?It is war between Islam and American infidels,? says Samee-ul-Haq, one of the main leaders of MMA, the religious political alliance in Pakistan. He was also the main supporter of the Talban regime.

While in the nineties, the American imperialism has left them on their own but not the major part of the establishment in Pakistan. They wanted to use them against their traditional rival Indians. They had also the illusions that a stable regime in Afghanistan would give them access to the markets of the Central Asia. So the formation and the support for Talban were the main priorities. All that changed with 11th September. The crushing of Talban in Afghanistan by bombing of US imperialism has not curbed the spread of religious fundamentalism. It has spread in a different form in Pakistan at present. Leon Trotsky, the architecture of the Soviet revolution of 1917 in Russia, once quoted ?You can not crush the will of the masses to change the society?. The will of the most of the masses to change the society has not emerged, as was the case in the past, through the left forces in the region.

Rather through the religious sentiment. The general elections in Pakistan held on 10th October gives some glimpse how they have emerged as a serious political force in Pakistan. The fundamentalist united in one main group called Mutehida Majlas Ammal MMA (United Action Forum) won the most seats in two provinces. They took considerable votes in all other constituencies. But the most alarming fact was their support in the two major cities, Lahore and Karachi.

At Lahore, they won three national assembly seats out of 13 and in Karachi, a neck-to-neck fight against the MQM, the immigrant mass force. MMA got 5 out of 20 seats in Karachi but it was very closely defeated in all the other 15 seats. MQM fetched 12 in the end. The MMA anti US rhetoric, which played a seminal role in wooing voters elsewhere in the country, played important part in their victory. But their campaign in Karachi was also centered on the issues like health and education. Mayer belonging to MMA already rules the local government in Karachi. MMA captured 29 out of 35 National and 52 out of 99 provincial assembly seats in North West Frontier Province (NWFP). In Baluchistan, it got 6 out of 13 National and 14 out of 51 provincial seat. In over all, MMA got nearly 3.2 millions votes. While the main bourgeoisie party MLQ got 7.3 millions. PPP fetched the most votes, nearly over 7.4 million votes. The total vote caste in Pakistan was over 28.9 millions. MMA got over 52 per cent of its seats in the low turnout areas. The fragmentation and disenchantment of the secular vote coupled with increased mobility and motivated election campaign by the MMA are definitely factors that cannot be ignored.

The conservative Muslim League, the main bourgeoisie party in Pakistan was split on the question of the support for military. A large part of it went to support the military rule while the rest under MLN supported the religious fundamentalist on the name of seat adjustment in many parts of Pakistan. MLN was taking revenge from the military as their leader Nawaz Sharif was ousted from power in October 1999. Later he was forced to leave Pakistan and now reside in Saudi Arabia. In the process, the MMA got more votes and seats than the MLN. MLN strategy made the fundamentalist stronger and is now no more close associates of the MLN. In fact MLN is fast becoming a tail end of the fanatics in Pakistan. MMA took more advantage of the lowering of the vote age from 21 to 18 than another political group. The Pakistan Talban from the Madrasas, the religious schools, over 70,000 of them in Pakistan, energetically worked for the successful campaign of the MMA. ? He who has the youth has the future,? Lenin said once. The majority of the youth is fed up with the system. "They are all corrupt", is the argument presented in most of the discussion among youth. ?Only Islamic revolution is the answer? you hear very often.

After the general election, the strategy of MMA was to give an impression to the masses of Pakistan that they are ready to shares the power but it has to be on their principal position. To attract and enlarge their support among the petty bourgeoisie, the MMA in words at least has become champions of the democracy and defendant of the 1973 constitutions.

The constitution, unanimously passed by parliament under Bhutto government, is a combination of capitalist democratic values and religious norms. The MMA also opposed the constitutional amendments by General Pervaiz Musharaf regime to give him more dictatorial powers as president. The Legal Framework Order (LFO) was announced on 21st August 2002, giving powers to president to sack the government anytime he likes and also powers for important postings in the establishment. This strategy has earned MMA some more respect and they are not seen as power hungry as was the impression of them in the past. The religious fanatics have supported most of the military regimes in Pakistan history. Only this time, they opted to remain in opposition in the center. They are more likely to form the provincial government in North West Frontier Province (NWFP) which is next to Afghanistan and could be part of the government in Baluchistan.

They have not gone to share the government at the center because of their perspective to become a mass force at national level. ? Had Punjab been awoke like the other province, the Islamic revolution in Pakistan would have been complete in this elections? cried Qazi Hussain Ahmed, the main leader of the most organized Jamaat-i-Islami; JI is a decisive component of MMA. It is JI successful political strategy of forming political alliance of all the religious forces that has won the mass support for the MMA. ?They are never united and can never be united? was the main arguments of the some of the Left intellectuals during the past years. MMA has at least made these intellectuals think twice before they bring this argument against them.

The unity process of the religious forces started long before the 11th September realizing the political vacuum after the failure of the successive civilian regimes to solve any of the basic problems of the masses in Pakistan. The 10 years of successive failed regimes of PPP and MLN after the fall of the last military regime in 1988, has provided the opportunity to the religious forces to exploit the opportunity.

In 1993, the JI formed a Pakistan Islamic Front with some minor Islamic groups to contest the general elections. The PIF failed badly and fetched less than 3 per cent of the total votes. In 1995, Mutehida Jakjehti Council (United Solidarity Council) was formed mainly to address the question of sectarian killings. In 2000, almost 29 religious groups formed the Islamic Mutehida Inqilabi Mehaz IMAM (Islamic United Revolutionary Front. This was mainly to deal with the Western ideas spread by the NGOs in the NWFP and Baluchistan. Jamiat Ulemai Islam under Fazal Rehman played the main role in this group. In June 2001, MMA was founded at the initiative of Jamaat-i-Islami. After 11th September, over 40 different religious groups formed Pak Afghan Defense Counsel. The MMA all eyes are on Punjab, the most populated province with over 64% of the total inhabitants living in this rather prosperous area of Pakistan. Many villages in Punjab particularly in Southern Punjab have young martyrs. Those who have been killed Kashmir or Afghanistan. These ?sacrifices? are laying a strong foundation for the possible growth of the religious political influence on those areas where MMQ is yet to become a majority. MMA activists unlike the main bourgeoisie parties are not driven by careers or greed. They are from the middle and working class background committed for a change. Most of the elected members of MMA from NWFP are from poor background and are seen by many all the time on cycles. This is contrary to elected members of rich parties who are on land cruisers and Mitsubishi Pajeros. The class question is very much in the mind of many who have gone to support the MMA in the recent elections. In the absence of the any significant left force in Pakistan MMA have echoed the class sentiments mixed with religion.

Would the American imperialism tolerate the growing influence of religious fundamentalism in Pakistan? The Americans has so for relied on the military generals to do the dirty job. Musharaf has done it very happily. But the political situations are getting out of control. The formation of the MMA government in NWFP will bring new contradictions between the new government in he center with the provincial government. The under ground Talban will feel easier in surfacing in NWFP under a religious government. Already new social tensions have emerged in the area. Women rights, human rights advocated in the province are under attack. ?NGO, s in these fields must wind up their offices? is the hidden message by the fanatics. The Sunday regular weekly off day will be once again replaced by Friday by the new provincial government. ?Beard is a must,? the new young elected leaders of MMA are advocating for every Muslim. ?Women must obey the Islamic norms and must cover themselves?. It is Talbanisation of Pakistani society in a different form. It is not yet advocating an open Jihad like the Alqaida did. But it is in the oven.

So NWFP and Pakistan could be the next target of all the propaganda of the US imperialism after Iraq. Any attempt by the center to dissolve or suspend the provincial government at NWFP will result in more civil war like situation. In 1974, the united government of the nationalist and religious fundamentalist was dissolved by Bhutto under the pressure of Shah if Iran, a guerilla war started in Baluchistan. Many were killed and many went to Afghanistan in exile. Now nationalist have diminished to large extent for the time being in this election, as they were seen as pro imperialists force. Many of these nationalists supported the American bombing in the false hope that it will crush the religious threat at home as well. More dark period is heading Pakistan unless new initiatives is been taken by the Left and progressive forces to stop the onslaught of the fanatics.

The handing over the recorded message of Asama Bin Laden in Islamabad to Algezeera tele network shows how things can develop in Pakistan. The possibility of the presence of Asama Bin Laden in Pakistan, the pro Talban provincial government in NWFP, the growing influence of religious fundamentalism in Punjab and Sind can force the newly elected pro Musharaf regime to take drastic actions against them. This in return will make the life of ordinary Pakistani more difficult. And it will make the ideas of religious fundamentalism more popular. Labour Party Pakistan had foreseen the recent growth of religious fundamentalism. It has declined to side with them on any issue. It has opposed the American bombing alongside with the condemnation of religious terrorism. It is adopting the policy to oppose the recently formed new civil pro Musharaf regime. The new civil regime is no change for the masses. It is the continuation of the Musharaf regime in civilian cloths. It will carry the IMF; World Bank dictated policies on the name of reform agenda. The religious fundamentalist will also oppose the reform agenda but only in words. Initially they are showing a very liberal face to the West. They have already said that privatization is a must. But under the pressure of the masses, they can make big noise against the economic exploitation of the people by the Multi nationals alongside with the social and political repressions. The hypocrisy of the fundamentalist can only be exposed with a united action of the trade unions, Left and progressive forces.

PAKISTAN AND JIHADIST VERSES

While Pakistan Army and civilian government are engaged with the US and other nations in fighting terrorism within their country as well as in Afghanistan, the volatile country is leaning more towards becoming a radical State than before. This is a dangerous trend particularly when the New York Times and Washington Post have reported that Pakistan is beefing up its nuclear arsenal at a breakneck speed and has already overtaken India in terms of the number of nuclear bombs at its disposal. Pakistan now has 110 nukes as compared to an estimated 60 to 80 nuclear bombs India is currently believed to have. Already Pakistan has emerged as the world's fastest growing nuclear arsenal power and is poised to displace the United Kingdom as the world's fifth largest nuclear weapon power.

India needs to take appropriate steps to deal with the national security threat the growing number of Pakistani nukes poses. The unpredictable, mercurial nation that India's Islamic neighbour is adds to New Delhi's problems. Unlike India, Pakistan does not subscribe to the no-first-use doctrine with regard to nuclear weapons. In fact, Pakistan had considered nuking India during the 1999 Kargil war. Therefore, it is all the more important for India in particular and the international community in general for Pakistan to behave responsibly and maturely. Unfortunately this is not happening and there are signs that Pakistan is hurtling headlong into the jihadist matrix. Two recent incidents are telling pointers to this grave development. In November 2010, for instance, Al-Rehmat Trust organised a series of courses on "jihadist verses' in Koran in more than 18 cities of Pakistan. The Trust is a front for Jaish-e-Mohammand, one of the most notorious terrorist groups operating from Pakistan with base in Bahawalpur and Peshawar. Jaish is headed by Maulana Masood Azhar, a former Harkat-ul Ansar rabble rouser, wanted in the 1999 Indian Airlines hijacking and several other terrorist attacks in India. According to the US Treasury Department, Jaish began using Al-Rehmat as a front organisation in 2002 after the former was banned by the US following the December 13 2001 attack on the Indian Parliament. The trust was used to recruit new set of terrorists for fighting the US forces in Afghanistan early 2009.

Another incident relates to October 2010 about the Taliban setting up schools on the outskirts of Karachi. These schools are outrightly terrorist schools where young students are taught bomb making and suicide bombings. One of the students, arrested by the police before he could blow himself up, said students were told that the Muslims in the world were being subjected to brutality and it was their duty to take revenge. It is not known how many schools and students are in these Taliban run schools. These are not the only Taliban schools. In fact, pro-Taliban madrasas have been in existence across Pakistan, including Islamabad, for quite some time. Lal Masjid in Islamabad was in fact one of the biggest of such madrasas claiming to work for turning Pakistan into an Islamic state and to establish Caliphate across the world through jihad. A survey conducted in Islamabad at that time showed Lal Masjid was not the only jihadi madrasa in the capital but at least 25 per cent of all mosques and madrasas either taught jihad or propagated extremist views.

These developments must be read with the support the Taliban has among the people, the army and the political leadership. One of the staunchest supporters of the Taliban is Pakistan's Tourism Minister Maulana Attaur Rehman. A Pashtun from Jamat Ulema-e-Islam-Fazlur Rehman party, Attaur said early this month at a public meeting that the Taliban were the true believers of Islamic ideology. He said it was the US which was the "biggest terrorist". There are others with equally strong views on the Taliban. These civilian leaders compliment the views held by terrorist leaders like Hafiz Saeed and Masood Azhar who leave no opportunity to propagate violence as a means to achieve whatever goals they have imagined for the Muslim world. Saeed has a much more powerful platform to propagate his terrorist views and objectives; he runs over 150 schools and more than 85 madrasas where hatred and violence are an integral part of the curriculum.

The school textbooks, published by his group, espouse a violent cause among the impressionable young students. His group also runs a powerful students body in Punjab University where its members promote extremist views among the student community and strongly oppose any attempt to project liberal views and opinions. It is not surprising that a large number of students from various colleges in Punjab find their way to terrorist training camps run by Lashkar-e-Toiba (LeT). In fact, a November 2010 report said several hundred students from Punjab were undergoing training at LeT camps in Pakistan Occupied Kashmir. Their agenda is to carry out terrorist attacks in India and to help the Taliban to fight the US-led forces in Afghanistan. With so many extremist forces brainwashing, indoctrinating and training the young students in Pakistan, its transformation into a radical state is not far. A deeply concerned India has to take pro-active measures to insulate it in case the Pakistani nukes fall into the hands of the jihadists by accident or by design of ever obliging elements in the Pakistani military and intelligence establishment. The increasing number of Pakistani nukes means more headaches for the Pakistani security establishment in safekeeping. It also means a higher success quotient for the terrorists in trying to steal the nuclear bombs.

EXPLAINING PAKISTAN'S COUNTERTERRORISM PERFORMANCE

The Afghan government's dissatisfaction and now increasingly the American polity's displeasure with Pakistan's performance in counterterrorism operations are conditioned considerably by the perception of Pakistan's unwillingness to crack down on terrorism comprehensively. This is a serious and, in actuality, complex charge. By all accounts, President Musharraf himself is strongly committed to purging both al- Qaeda and the Taliban. The imperatives of eliminating al- Qaeda are obvious: Pakistan was never directly a sponsor of this group in Afghanistan, and destroying its network remains the sine qua non of the lucrative Pakistani counterterrorism partnership with the United States. Musharraf also remains personally opposed to the political philosophy represented by the Taliban. He has repeatedly identified the "talibanization" of Pakistan as the most pressing threat facing his state, but whether this translates into a decision to physically apprehend or eradicate the Taliban cadres, and especially their leadership, is less clear. Drawing a distinction between "diehard militants and fanatics," who "reject reconciliation and peace" and accordingly must be targeted even though they are hard to find, and the larger Taliban cadres, "most of [whom] may be ignorant and misguided" but "are a part of Afghan society," Musharraf has urged Kabul, Washington, and the larger international community to begin instead a campaign of reconciliation with the Taliban focused on "winning [their] hearts and minds." Musharraf's attitude toward the Taliban thus remains complex and multifaceted: he clearly detests their worldview, referring to talibanization as a species of extremism that "represents a state of mind and requires [a] more comprehensive, long- term strategy where military action must be combined with a political approach and socioeconomic development." He is also opposed to what he calls "terrorist elements and foreign militants" within the movement, which he acknowledges "must be dealt with a strong hand." Musharraf argues, however, for peacefully integrating the Taliban's rank and file into civil society, a judgment that is premised heavily on the belief that these elements are merely misguided miscreants rather than implacable foes.

The nucleus of military officials around Musharraf appears to reflect his own sinuosity. While all senior Pakistani military officers are agreed that the al- Qaeda presence in the FATA must be eliminated, there is a considerable diversity of views in regard to the Taliban. Although many feel that the optimal outcome for Pakistan would be simply a Taliban that progressively lose their effectiveness and support and thereby fade into obscurity— a finale that would spare Pakistanis the distasteful obligation of having to turn their guns against their old clients— others are conflicted about these reactionaries for different reasons.

To begin with, many officers are disenchanted by Washington's approach to managing the larger issues associated with Afghanistan's political reconstitution. Since the emergency loya jirga held in 2002 and the subsequent Afghan presidential election of 2004, these officials have been dismayed by what they perceive as the U.S. partiality toward the Durrani Pashtuns, who have traditionally been the privileged political elite in Afghan society (and from whose ranks, through the Popalzai tribe, emerged President Hamid Karzai). The calculated neglect of the Ghilzai Pashtuns— who are primarily rural and uneducated peasantry and who constituted a critical source of manpower for the Taliban cause— grates on many Pakistani national security managers not only because they believe that the continuing alienation of the Ghilzai feeds the Taliban ranks but also because it represents an enduring and deliberate disregard of their own clients in intra-Afghan politics. This latter consideration is significant because it has the effect of portraying Islamabad as feckless and incapable of influencing U.S. policy in directions more helpful to its friends despite Pakistan's large investments in the U.S. war on terror.

When these concerns are added to other strategic calculations about protecting the Taliban as a hedge against either the failure of the Karzai regime in Kabul or the dreaded prospect of increasing Indian influence in Afghanistan, the senior leadership of the Pakistani military— as well as President Musharraf— believe they have good enough reasons to avoid targeting the Taliban comprehensively in the manner sought by both Kabul and Washington. The dangers of a heightened and targeted anti- Ghilzai campaign leading to a political mobilization that renews the demand for an independent "Pakhtunistan" further exacerbate the fears of senior Pakistani military officials. Because the status of the Durand Line separating Afghanistan and Pakistan is still formally contested by Kabul, any provocation that results in strengthening the political dimensions of Pashtun solidarity among the tribes living on both sides of this boundary is viewed immediately as a potential threat to the territorial integrity of Pakistan. With the vivisection of 1971 indelibly emblazoned in the consciousness of the Pakistani military, senior commanders are extremely reluctant to embark on any military operations that would aggravate the local Pashtun tribes in the FATA and provoke them into making common territorial cause with their confreres on the other side of the border against the Pakistani state.

While ambivalence about the Taliban at senior levels in the Pakistani military thus has both strategic and self- interested dimensions, other more prosaic, but tactically important, considerations also play a role. Recognizing that the Taliban are essentially Ghilzai Pashtuns with deep consanguineal ties to the tribes that have dominated the FATA for centuries, many Pakistani commanders are afraid that any continued large- scale military presence in the area, especially if exemplified by massed infantry operations of the kind mounted in 2002–2004, will only further inflame tribal sensitivities and diminish cooperation between tribal leaders and the armed forces, a cooperation that is absolutely necessary if the armed forces are to successfully apprehend the "terrorist elements and foreign militants" located in their midst.

Although such embitterment has already occurred with problematic consequences for the success of antiterrorism operations, the leadership of the Pakistani military remains continually fearful that any added aggravation could lead to even greater tribal support for terrorist groups closeted in the area and a systematic denial of access to the military units tasked for operations therein. That such contingencies already appear to have materialized in the FATA suggests that the reluctance of senior Pakistani military leaders to violently engage the Taliban will be reinforced even further.

The most problematic elements within the Pakistani state, however, are probably the ISID officers in the field who were tasked with managing the liaison relationship with the Taliban over the years. Some simply feel loyalty to their old clients. Others are content to exploit their leadership's own ambivalence about the Taliban. And some others are prepared to disregard leadership directives that enjoin interdicting the Taliban for either nationalist, ideological, or personal reasons— if they believe they can get away with it. Whatever the cause, the field operatives of the ISID are widely perceived in Afghanistan and in the United States as being less than fully committed to targeting the Taliban leadership in the manner required for the success of counterterrorism operations in the FATA and beyond.

At first sight, this is indeed a curious phenomenon because nothing in the organizational structure of the ISID suggests that it is either an autonomous or a rogue entity. The reportedly 10,000-strong ISID is staffed primarily by Pakistani military officers who are assigned to the service on deputation for a fixed period of time, and its leadership reports to the chief of army staff. The pay, promotions, and operations of the directorate are also regulated by military rules and procedures, and by all accounts the Pakistan Army is a professional and bureaucratically efficient organization. Consequently, the notion that ISID officers might be undermining policies pursued by the corporate leadership of the Pakistan Army appears counterintuitive at first sight and cannot be reconciled with the image of the Pakistan Army as a tightly centralized organization unless due credit is given to three realities.

To begin with, the ISID, similar to many other intelligence organizations worldwide, has considerable operational latitude because of the nature of its activities in the covert realm: this includes access to financing "off the books," recruitment of agents from diverse sources to include those with unsavory backgrounds, and the systematic use of retired case officers who can conduct officially permitted operations while still providing the state with plausible deniability.

Further, the implementation of many ISID operations is typically regulated by "directive control" as opposed to "detailed control," where field officers have the flexibility to accomplish strategic goals without having to secure prior approval of every particular from their superiors. And, finally, because ISID is simultaneously an external intelligence organization as well as a coercive instrument for implementing the preferences of military authoritarianism in Pakistan, Musharraf's management of this organization historically was manifested primarily through the promulgation of broadly defined policies, which were then implemented by a chain of subordinates who acted upon their understanding of his strategic intent and which simultaneously served to protect him from detailed knowledge of what may frequently have been highly troublesome activities.

Given these realities, the pervasive belief about ISID unreliability in the war against terror, and particularly against the Taliban, can be accounted for— despite the otherwise professional character of the Pakistan Army— only by one or more of the following hypotheses:

- That, despite their public claims, President Musharraf and his corps commanders are not yet committed to a policy of eliminating the Taliban and especially its leadership, root and branch; consequently, Musharraf and his commanders have not directed their military and intelligence services to systematically implement such a strategy.

- That, although President Musharraf and his corps commanders have settled on a strategy of eliminating the Taliban in principle, the operational predicates of this policy insofar as they apply to the leadership and other high-value targets have not yet been specified deliberately, thus permitting line- level officers to use their discretion when it comes to supporting or undermining particular counterterrorism operations.
- That, although President Musharraf and his corps commanders have settled on a strategy of eliminating the Taliban, to include its leadership and other "diehard militants," the large size and complex bureaucratic structure of the ISID permit its field officers, and especially the retired case officers still on its active payroll, to covertly ignore or violate leadership directives in many instances for nationalist, ideological, or personal reasons without fear of immediate retribution.

Whatever the real explanation for ISID recalcitrance may be—and the truth probably implicates a complex admixture of factors—the fact remains that the Pakistani campaign against the Taliban, and particularly its leadership, remains hobbled by convolution and hesitation. During the past several years, this has resulted in a deepening entrenchment of the Taliban and their sympathizers throughout the seven administrative agencies of the FATA and a growing expansion of their influence in the Tank, Dera Ismail Khan, and Swat Valley areas of the North West Frontier Province of Pakistan. Since October 2007, for instance, the idyllic mountain region of Swat, barely 90 miles from Islamabad, has been occupied by self- styled "Pakistani Taliban" forces led by Maulana Qazi Fazlullah and his Tehrik- e- Nafaz- e- Shariate- Muhammad (Movement for the Enforcement of Islamic Laws) shaheen (fighters), who may be present in the area in greater than brigade strength. This development suggests not only that the Taliban movement and its sympathizers have moved beyond the traditionally stateless regions close to the Durand Line and into more settled areas within Pakistan but also, and more ominously, that what began as localized terrorist operations now threatens to evolve into a mature insurgency with the militant opposition able to eject government forces from a given territory, hold ground against state opposition, and coerce any local opponents into cooperating in order to sustain the newly secured safe haven.

This intensifying talibanization in the sensitive areas of the North West Frontier Province has had diverse effects, including increased tensions with and between the traditional tribes resulting in both growing intertribal conflicts as well as bolder attacks on the Pakistan Army and paramilitary units in the region.

In one dramatic encounter, more than three hundred Frontier Corps infantry men were taken hostage by local militants in August 2007 in South Waziristan, to be freed only after President Musharraf ignominiously released more than two dozen previously jailed Islamists, including Mullah Obaidullah Akhund, the highest-ranking Taliban official captured by the Pakistani military. On occasion, the tribal dissatisfaction with local talibanization, and particularly with some of the groups that support it, has been all to the good as, for example, when indigenous tribes in the Azam Warsak area of South Waziristan, supported by Pakistani military forces, attacked and expelled numerous Uzbeks belonging to the Islamic Movement of Uzbekistan— a constituent of Osama bin Laden's International Islamic Front— who had settled in that locale. Unfortunately, such successful cleansing operations have been all too few. The main consequence of talibanization in the FATA instead has been, as one observer put it, to "provide more opportunities to the ISI[D] to indirectly support some Taliban commanders sympathetic to Pakistan's objectives" in the ongoing war in Afghanistan.

The growing talibanization in the FATA and beyond has in fact resulted in the creation of a secure sanctuary for a variety of terrorist groups now conducting anticoalition military operations in Afghanistan. The evidence suggests that Taliban presence is strongest in the Helmand, Kandahar, Zabol, and Oruzgan provinces in southern Afghanistan and is either significant or conspicuous in Paktika, Khowst, Nangarhar, Konar, and Nuristan provinces in eastern Afghanistan. In these areas, the Taliban have been able to deploy and sustain a large number of armed fighters in situ, which has permitted the movement to effectively displace the Afghan state by usurping its traditional functions such as maintaining law and order; extracting resources through taxation; administering justice through various adjudicative mechanisms backed by local militias; and dispensing welfare through the maintenance of schools, provision of social services, and the oversight of economic activities.

Performing these statal functions is possible because the Taliban undoubtedly continue to derive support from the Ghilzai Pashtuns in Afghanistan; however, their record of increasingly ruthless retribution against any uncooperative tribal leaders does not hurt either. In this context, the Taliban's ability to sustain the strong coercive presence they currently have in the southern and eastern Afghan provinces is enhanced considerably by their access to the safe haven in the FATA whence they can draw a large number of fighters, procure diverse kinds of ordnance and combat equipment, and tap into different streams of financial resources (including, but not restricted to, the zakat, the Islamic tithe) for the prosecution of the ongoing jihad against the Karzai regime and the foreign forces now present in Afghanistan. As one respected analysis has noted, "using [the sanctuary provided by the FATA] to regroup, reorganize and rearm," Taliban and other foreign militants, including al- Qaeda sympathizers, "are launching increasingly severe cross- border attacks on Afghan and international military personnel, with the support and active involvement of Pakistani militants."

The fruits of this activity are witnessed in the deteriorating security situation in Afghanistan. Since the successful presidential election in October 2004—an event that received the full support of the international community and heralded hope for a new Afghanistan— the Taliban insurgency has metastasized in scale, intensity, and fury. By 2006, the level of violence had increased dramatically, and previous operations that had been centered on assassinations, ambushes, and isolated hit- and- run attacks were now supplemented by more ominous tactics involving beheadings and suicide bombings, which historically have been utterly alien to Afghan culture. Equally problematic has been the employment of ever more sophisticated improvised explosive devices, rockets, missiles, and man- portable air defense weapons, many of which continue to be fabricated in the arms foundries in the FATA but are now also increasingly supplied by al- Qaeda and foreign powers such as Iran. Even more remarkably, Taliban military operations have gradually evolved from singular covert attacks mounted by tiny groups to more complex, set- piece military operations undertaken by larger units, often involving attempts to seize and hold territory against superior forces, and employing more diverse and sophisticated crew- served weapcnry, including indirect fire systems such as mortars and unguided rockets, as supplements to the traditionally ubiquitous personal firearms and direct- fire weapons.

The strategic objectives of these new modes of warfare have also become more complex. Rather than simply harassing the new Afghan government, which seemed to be the original intention, the current military activities of the Taliban are accompanied by sophisticated forms of information operations. Betraying evidence of lessons learned from their al- Qaeda accomplices, Taliban operatives use a variety of techniques ranging from sending crude low-technology "night letters" often conveying threats to specific individuals, to circulating DVDs and videotapes containing political propaganda, to exploiting more advanced technology such as radio, television, mobile and satellite telephony, and the Internet. In general, all these technologies are used to signal to the Afghan tribes that the return of the Taliban to power in Kabul is inevitable— despite whatever tactical losses might be suffered at the hands of NATO forces in the interim— and that resistance or neutrality is therefore futile. As one analysis pointed out, even if this campaign does not persuade the Afghan people, "[t]he Taliban's own hearts and minds activities are now prolonging and exacerbating an already difficult insurgency problem for the Afghan Government and the International Security Assistance Force (ISAF) in the south of the country."

Finally, and perhaps most important, both the Taliban's information and its actual war-fighting operations have moved beyond simple terrorist attacks aimed at disrupting the Afghan government toward more ambitious objectives revolving around the progressive domination of territory. Focused today primarily on NATO's Regional Command (South) for both symbolic and strategic reasons, the Taliban leadership appears intent on slowly seizing critical areas, district by district, through a strategy of covert infiltration in the Helmand, Kandahar, Oruzgan, and Zabol provinces as a prelude to wresting control of the city of Kandahar, which is intended to become the base for first dominating the South and eventually all of Afghanistan itself. If this evolution gradually succeeds, the Taliban insurgency in southeastern Afghanistan will have successfully metamorphosed from a guerrilla operation into something resembling a more conventional civil war with grave advantages to the militants in their struggle against the Karzai government in Kabul.

Although the safe havens in the FATA and the ability to derive local Ghilzai support in the southern and eastern Afghan provinces (by providing the benefits of security, justice, and even development) have thus enabled a dramatic transformation in both the character of the ongoing Afghan war and the fortunes of the Taliban as an insurgent organization, a particularly dangerous consequence— especially from a U.S. perspective— has been the enhanced prospects for survival they have offered al- Qaeda. There is little doubt today that the survival of the Taliban sanctuary in the FATA (to include the talibanization of the wider area more generally) has been singularly responsible for the continuing regeneration of al- Qaeda as an organization because it has permitted the leadership and the operatives of this terrorist group, who are relatively smaller in number, to safely "dissolve" into a larger geosocial environment that is either hospitable to them directly or that protects them by disguising their presence amid a larger pool of Taliban adherents.

The al- Qaeda leadership, which is believed to be currently ensconced somewhere in the Bajaur Agency of the FATA, has further enhanced its immunity to interdiction by pursuing what appears to be a subtle strategy toward its Taliban hosts. Recognizing that the Taliban's Pashtun cadres remain the original denizens of the FATA and the adjacent areas in eastern and southern

Afghanistan, al- Qaeda's overseers have been careful to tread lightly: despite their independent access to significant streams of foreign resources, they do not seem to have levied any excessive demands in terms of either hospitality or security, nor have they used their superior access to advanced military- technical capabilities worldwide to attempt any "takeover" of the Taliban movement. Rather, they appear to understand that an independent Pashtun insurgency that answers to no one but its own indigenous leadership stands the best chance of not only regaining control in Afghanistan but also securing the continued support of the tribal elements in the FATA, which in turn only better conduces to al-Qaeda's survival over the long term.

Al- Qaeda leaders thus have repeatedly endorsed the Pashtun leadership of the Taliban, centered on Mullah Omar's coterie, on many an occasion publicly, beginning in 2002 when Osama bin Laden conferred on Mullah Omar the title of Emir Al- Momineen (Leader of the Faithful). The increasing sophistication of the Taliban's military operations, the new integration of suicide attacks into its modus operandi, and its increasing emphasis on information operations for a group that historically despised the modern media also indicate that al- Qaeda elements continue to assist Taliban forces with at least technology and training, and possibly financial assistance, as partial recompense for the refuge they receive in the FATA as they continue to bide their time awaiting the reestablishment of Taliban control in Afghanistan.

Pakistan's failure to target the Taliban and especially its leadership since 2001 has, therefore, had several deleterious consequences. To begin with, it has resulted in the creation of a safe haven for various terrorist elements in the FATA, whence the Taliban war against the Karzai regime can be prosecuted and the al- Qaeda leadership protected and regenerated as it plans more catastrophic attacks on the West and on the United States in particular. It has also permitted the Taliban to nurture their indigenous bases of support within southern and eastern Afghanistan itself, whence they can slowly evolve into a tumorous state within a state. Further, it has bred a cancerous nest of violent extremism inside Pakistan resulting in the rise of new Islamist militant groups, sometimes labeled the Pakistani Taliban, that are either sympathetic to or affiliated with al- Qaeda and committed to waging a holy war against the Pakistani government, the liberal elements in Pakistani politics, as well as other foreign adversaries such as India, Israel, and the United States. The invigoration of these indigenous radical outfits has in the process produced a new generation of foot soldiers available to different extremist entities throughout the country and strengthened the social bases of support for the otherwise marginal Islamist parties in Pakistani politics. Finally, it has added to the already long and intractable list of problems confronting Pakistan as it struggles to transform itself into a moderate and successful Muslim state: in particular, it has condemned the Pakistani leadership, including acknowledged moderate leaders like Musharraf, to prosecute antiterrorism operations under highly disadvantageous conditions and in an area that by history and tradition has long been lawless, has been bereft of any concentrated state penetration, and that had no regular military presence worth the name until recently, yet is dominated by those very groups that have strong ethnic and increasingly ideological ties to the same terrorist elements sought by the Pakistani state.

If the foregoing discussion amplifies how Pakistan's counterterrorism performance has been structurally compromised by motivational and institutional problems, this is by no means the whole story. An equally important source of inadequacy has been the operational complexity of the counterterrorism operations themselves and Pakistan's myriad weaknesses in coping with these challenges. These difficulties— three of which are illustrated in the discussion that follows— only complicate the challenges caused by the larger problem of whether Pakistan believes eliminating the Taliban decisively is in its national interest.

First, Pakistan's inability to secure the tactical intelligence required for successful counterterrorism operations against key Taliban and al- Qaeda elements in the FATA has now become painfully obvious. Although the ISID and the army's director general of military intelligence have primary responsibility for the collection of targeting intelligence in the FATA, their ability to carry out these tasks has been severely hampered in recent times. In part, this is undoubtedly because many Pakistani intelligence officers are simply sympathetic to radical Islamist elements who have been their clients for many years. Even when this is not the case, however, state intelligence activities have been hindered by the peculiarities of the political structures in the FATA and the corrosive changes that have been occurring therein.

It is often insufficiently recognized that, although the tribal areas are physically located within Pakistani territory, they are not governed by either Pakistani laws and regulations or the political institutions normally associated with national politics. In fact, the relationship between these tribal areas and the Pakistani state is regulated not by any common laws but by formal treaties between the resident tribes and the federal government in Islamabad. The existence of such treaties exemplifies what two analysts have rightly labeled "the anomaly of [the] FATA": it signifies that the link between the tribes and the Pakistani government resembles one that exists between coordinate, and not superordinate and subordinate, political entities. This is further corroborated by the fact that these treaties not only guarantee the tribes' immunity to the codified laws and regulations that govern political life in the rest of Pakistan but also bestow on them exclusive responsibility for the management of their own internal affairs. With the exception of the Frontier Crimes Regulation, a written document more than a century old that elaborates the principle of settling disputes through arbitration by tribal jirgas, most of the governing rules in the FATA are essentially unwritten, being based on a combination of rewaj (tribal customs) and Sharia (Islamic law).

The foundation of maintaining order and authority in such a system, which is anchored in custom, tradition, and legal practices going back to the British Raj, lay in the inculcation of harmonious relations between the political agent— a mid- level civil servant with sweeping powers who, although deputed by the governor of the North West Frontier Province as the highest- ranking official representative in each tribal agency, was ultimately responsible to the federal government in Islamabad— and the tribal maliks, or elders, who managed tribal affairs day to day and who until 1996 were the only individuals permitted to vote in elections for Pakistan's National Assembly. The ties between the agents and the maliks were critical to the production of good intelligence: the agents disbursed the resources provided by Islamabad to acquire the information required to keep the government up to date about developments along the frontier, and the maliks used the subventions provided to buttress their own influence, access, and standing with the tribes they supervised.

Although it was possible to alter this traditional structure of management in the FATA, successive authoritarian regimes in Islamabad eschewed that alternative because the system of direct control through the political agent invariably appeared more attractive to Pakistan Army leaders who were innately uncomfortable with the idea of democratic alternatives involving the introduction of universal adult suffrage, the development of representative institutions, and the presence of civilian political parties in local politics. As a result, the traditional governing mechanisms, centered on the interactions between agents and maliks against the backdrop of the privileges encoded in the old treaties, were only reinforced by Islamabad despite the fact that the bonds between these agents and maliks had became increasingly discredited because of the widespread corruption and politicization that came to characterize their relationship. As a result over time, the tribes along the frontier no longer looked up to their own maliks as selfless leaders or to the political agents as fair representatives of a federal government that sought to advance their welfare.

By the time the 1980s set in, the anti- Soviet jihad brought about a further— and deadlier— acceleration of this crisis. Egged on by the initiatives of Pakistan's Islamist president, Zia ul- Haq, the FATA witnessed a steady social transformation that resulted in the traditional authorities— the political agents and the maliks— being slowly supplanted by new religious leaders, the maulvis, who viewed issues of political loyalty primarily through religious or ideological lenses. The progressive demise of the old social order thus made the long- standing Pakistani human intelligence collection apparatus dramatically ineffective as the radicalized maulvis, viewing the protection of the Taliban and al- Qaeda cadres in the FATA as a politico- religious obligation, appear determined to deny the Pakistani state the necessary information required to apprehend these targets. The widespread outcry in the frontier areas against the U.S. war in Iraq, coupled with the growing perception that Musharraf's prosecution of counterterrorism operations represents illegitimate support for a U.S. administration involved in a global anti- Muslim crusade, has only strengthened the determination of the maulvis and the new Islamists, who have filled the "power vacuum" caused by the demise of the agent-malik relationship in the FATA, to protect the terrorist targets sought by Pakistan and the United States.

The limitations of Pakistani technical intelligence capabilities in the context of counter terrorism operations also do not help matters any. As a matter of fact, Pakistan does have an impressive array of national intelligence collection capabilities. These systems, which are focused primarily on gathering signals and communications intelligence (SIGINT and COMINT), are largely under ISID control although the actual intercept operations are conducted by inter- services signals units that employ technical personnel drawn from the army's Corps of Signals, the air force, and the navy. For the most part, however, strategic SIGINT and COMINT collection in Pakistan— the intercept, analysis, and dissemination of electronic signatures and communications waveforms— is disproportionately oriented toward targeting India. Islamabad's most sophisticated assets, accordingly, focus on the detection, direction finding, surveillance, and intercept of the high frequency (HF), very high frequency (VHF), ultra high frequency (UHF), and satellite bandwidths used by Indian diplomatic and military communications. These resources, together with the tactical SIGINT and COMINT systems possessed at the corps level in the Pakistan Army, make Islamabad certainly capable of monitoring the communication devices used by the Taliban and al- Qaeda, because these in fact most likely resemble those supplied by the ISID to various Kashmiri terrorist groups and recovered over the years by the Indian military. The insurgency in Kashmir revealed that Pakistani-supported terrorist groups in South Asia, including those operating in Afghanistan, generally use HF radio, satellite telephony, and cellular phones for long- range connectivity, with commercially available line- of- sight VHF and UHF radios produced by companies such as Yeasu, Kenwood, and I- Com for their operational and tactical communications.

Targeting the communications traffic generated through these systems, however, requires Pakistan's national and tactical collection assets to be systematically tasked for this purpose, but both Indian and Afghan military intelligence officials believe that New Delhi continues to remain a higher- priority target for Pakistani technical collection in comparison with either the Taliban or al- Qaeda. Even when this is not the case, however, Pakistani surveillance systems may continue to be ineffective in the counterterrorism mission for many reasons. If Taliban and al- Qaeda operatives use low- power devices sporadically for tactical communications, the short range and random nature of these transmissions may defeat even a technically competent operator if no surveillance devices are in proximity to the threat. Further, sophisticated technologies such as frequency hopping, portable encrypted, or digital burst radios, many of which are available commercially, can be used to elude even skilled surveillance especially if the monitoring systems are not available or are not dedicated full- time to the mission.

Finally, the increased use of the Internet by Taliban and al-Qaeda operatives, including their growing use of encryption software, makes it hard for the ISID to monitor such communications systematically because, in the absence of prior cueing, high- speed computation married to sophisticated search algorithms would be required if the relatively large volume of Internet traffic, even within an otherwise relatively low tele- density state like Pakistan, is to be successfully monitored. It is simply not clear whether Pakistan possesses such capabilities.

In principle, U.S. advantages here could serve to compensate, but the growing appreciation of the capabilities of U.S. assets has resulted in these opposition forces— both Taliban and al- Qaeda— increasingly relying on more primitive but more secure means of communication, such as "snail mail" and human couriers, for their operational planning. This workaround, in turn, denies both

Pakistan and the United States the kind of targeting data that might otherwise have become available through technical intelligence.

Recognizing these problems, Pakistan has begun the arduous task of rebuilding both its technical and its human intelligence collection assets in the FATA. The latter capabilities are indeed the most critical, but these also take the longest to mature and to yield their fruit. A long- term Pakistan Army presence in the FATA amid conditions of relative peace is, therefore, an essential precondition for Islamabad to be able to develop and consolidate an effective human intelligence network. The $750 million U.S. assistance program to the FATA, if properly directed, could help considerably in advancing this goal of local stability; but the complicated and time- consuming nature of this endeavor, the uncertainty about the program's effective implementation, and Washington's failure to condition the availability of these funds on Islamabad's implementation of political reforms in the tribal regions— to include, inter alia, the drastic revision of the Frontier Crimes Regulation; the elimination of the political agent as part of the larger process of integrating the FATA into Pakistan's North West Frontier Province under the full jurisdiction of the provincial and national legislatures and the judicial system; and the withdrawal of restrictions on political parties operating in the FATA with an eye to introducing conventional political institutions— imply that neither the United States nor Pakistan ought to expect quick breakthroughs in their efforts "to build confidence and trust between the Government of Pakistan (GOP) and [the] FATA tribal communities" leading to the demolition of the al- Qaeda and Taliban networks that have regenerated in this area over the last few years.

Second, the arrival of the Pakistan Army in strength in the FATA has resulted in social disruptions that have undermined its counterterrorism effectiveness. Although the insertion of the Army's XI Corps and the SSG battalions into the autonomous areas was a brave and necessary decision of the part of General Musharraf, it has nonetheless eroded the delicate compact that previously existed between the FATA and the Pakistani state. The resulting alienation and resentment on the part of the indigenous population have been reflected in significant counterterrorism problems. The Pakistan Army— which draws its cadres largely from outside the FATA and is primarily non- Pashtun in composition— is a highly professional force, but its maneuver units have often been stymied by their inability to secure the cooperation of the local populace, which views it today as an unwelcome intruder. The army's SSG is very effective in tactical counterterrorism operations but, being an elite unit, is far too diminutive to make a difference at the theater level.

The Frontier Corps, which is composed primarily of tribal levies and is the resident paramilitary force, could be potentially the most effective element, but it is often compromised by its close ties with the local inhabitants. Riddled with sympathizers, inadequately motivated, suspicious of both Islamabad's and Washington's intentions, poorly trained and equipped for counterterrorism operations, yet present in strength throughout the FATA, the Frontier Corps (along with its other local siblings such as the Frontier Constabulary, the tribal police [khassadars], and tribal militias [lashkars]) represents the perfect exemplar of the structural challenge facing Pakistan's counterterrorism effort: its best local units, the ones that share affinities with the tribes they patrol and consequently the forces likely to secure potentially the most useful intelligence, are also the fighting elements least able or willing to cope with the battle-hardened terrorists they are deployed against.

Unfortunately, the infantry elements of the Pakistan Army that have been pressed into the fight have their own problems as well. Unlike the Indian Army, which thanks to two decades of combating Pakistani- supported subconventional conflict now has considerable counter terrorism skills, the infantry battalions of Pakistan's XI Corps are configured primarily as strategic reserves for possible conventional warfare against India. Counterterrorism operations are not their forte, and, while they have done a decent job of learning by doing, they still betray a proclivity for operational responses that while sensible against a conventional adversary are less than effective (and, perhaps, even counterproductive) when dealing with irregular forces: large unit deployments, intense (and sometimes indiscriminate) employment of fire, and sledgehammer cordon- and- search tactics.

While the attrition strategies of the Pakistani military have been criticized by many for their detrimental consequences, it must be recognized that these are not always attributable to the "self- proclaimed invincibility of the [Pakistani] armed forces." Rather, the hostile terrain in which counterterrorism operations are conducted and the unexpectedly heavy firepower that Taliban and al- Qaeda terrorists have mustered in the past have been the two factors most responsible for the military's recourse to the relatively coarse counterterrorism tactics that are invariably derided. It is not difficult to sympathize with Islamabad's predicament. For starters, the topography of the FATA is incredibly inhospitable as far as counterterrorism operations are concerned. The general geography of the area is characterized by harsh, rugged, and inaccessible mountainous terrain with steep slopes being the rule rather than the exception. The crest elevations in the region vary from 3,600 meters to 4,700 meters in the Khyber, Kurram, and Orakzai agencies of the central FATA, dropping somewhat to between 1,500 meters and 3,400 meters in the southern agencies of North and South Waziristan. These mountain ranges running roughly from northeast to southwest function as a complex barrier that breaks up the terrain into numerous tiny basins or valleys that are dotted with minuscule settlements surviving either through livestock grazing, subsistence agriculture, or petty trade. The size of these settlements is generally very small, ranging from literally a few dozen people in some instances to a few thousand at most in the largest hamlets.

The lines of communication between these outposts are invariably tenuous, extending along the ridgelines of the adjacent mountains or traversing them through numerous passes, tracks, and trails, many of which support only pedestrian traffic or pack animals. Because many of these routes are intestinal and insignificant, they are often known only to the locals who, along with smugglers, drug runners, and arms peddlers, have exploited these conduits in the natural terrain to carry out their business undisturbed for centuries. These terrain features produce three significant operational consequences that have great impact on the conduct of military operations.

First, the isolation of the hamlets amid craggy geophysical features and the small sizes of the populations sheltered within them make it virtually impossible for outsiders to monitor any personnel movement to or from these locations, especially if the transit occurs on foot, by animal, or by isolated vehicular traffic (where possible). This is especially true if the movement concerned occurs in adverse weather or at night. Second, the consanguineal character of the tribal populations living in these areas implies that strangers cannot travel within the area without being readily detected, and safe passage in such circumstances usually occurs only when the local inhabitants are persuaded about the alien's peaceful intentions through some form of attestation by individuals known to the resident tribes. Third, the distances between the populated outposts can be significant given the absence of paved or metaled roads and, consequently, quick movement across the terrain invariably requires either strenuous marches on foot on or off established paths (depending on circumstances) or the use of animals, accompanied by guides in most cases. In several locations vehicular traffic is in fact possible, but, because such movement invariably hews to well- established roads and pathways, covert entry and exit through such access routes is generally difficult.

This concatenation of features abundantly explains why Pakistani counterterrorism operations have often run into tactical difficulties requiring recourse to "excessive" force. The isolated setting of many FATA settlements where terrorist cadres find refuge essentially prevents the Pakistani pursuers of the terrorists from being able to approach these locales clandestinely. Even small commando units operating on foot are susceptible to premature detection by the locals, and the munitions and weapons required to be carried over the harsh terrain and along the great distances within the region often tax the abilities of even the fittest infantry units, which must conserve their strength for the arduous military action at the end of their insertion. This consideration invariably mandates traveling on established tracks and paths, but even off- track approaches do not provide any assurance that the attacking force will be able to close in on its target undetected.

Because the risk of compromise is consistently high, many terrorist refugees have been able to simply escape at the first warning of military units moving en route to their hideouts. Early engagements with the Taliban and al- Qaeda cadres who chose to remain bivouacked also revealed— often to the surprise of their attackers— just how heavily armed they were; in fact, the character of their military equipment could often make the difference in whether they chose to escape or stand their ground and fight. Whenever they settled upon fighting, their employment of heavy weapons was invariably made doubly effective by the natural advantages accruing to the defense especially in mountainous terrain— gains that were further magnified by the clever use of stealthy tactics, the cunning utilization of the surrounding topography, the erection of effective positional defenses, and the exploitation of the timely warning provided by the local inhabitants.

The persistence of such challenges compelled the Pakistani military to seek operational work- arounds that offered some chance of success. The solution that proved most attractive in many circumstances was to forgo tactical surprise, which might have ensued from the exclusive use of small units relying entirely on covert foot penetration, in favor of larger operations that sought to exploit tactical superiority through the employment of heliborne elements for both the transportation of substantial strike teams to some location in proximity to their designated target and for the firepower required in support of the actual assault. Because the final engagement in such situations usually involved a heavily armed and a partially or fully alerted adversary— if the latter had not already escaped— the attacking Pakistani combat teams were often forced to employ even heavier weapons than might have been originally intended, including mortars, antitank recoilless rifles and guided missiles, field artillery, helicopter- and aircraft-fired cannon and unguided rockets, and occasionally even general-purpose bombs delivered by tactical aircraft. The lessons offered by such engagements since 2002 are stark and clear: unless the tribal populations residing in the FATA are sympathetic to the government and are willing to either warn the army of the militants' presence in their midst or desist from alerting the terrorists to the military's anticipated arrival in their hamlets, counterterrorism missions will either fail or be condemned to rely on even greater applications of brute force for their success.

The inevitable, but unintended, consequence of implementing such solutions has been significant collateral damage among civilians in the tribal areas. The residents, in response, have reacted to these losses by mounting violent attacks on, and repeated seizures of, Pakistani troops and paramilitary forces deployed in the area. The more extremist outfits, to include al-Qaeda elements, have sought to exact their revenge by undertaking lethal suicide attacks against Pakistani military and intelligence personnel both within the FATA and deep inside the nation's heartland in an effort to compel President Musharraf to terminate his counterterrorism operations conclusively. These continuing attacks on Pakistani military personnel have, by many anecdotal accounts, lowered morale within the frontline units now operating inside the FATA and caused increased desertions, suicides, and frequent discharge applications. Not surprisingly then, these developments have induced deepened soul- searching on the part of local commanders who wonder about the strategic wisdom of the ongoing war on terror and question the benefits specifically accruing to Pakistan. The growing antagonism caused by the collateral damage associated with U.S. military strikes from the Afghan side of the FATA has not helped make the Pakistan Army's problems any easier in this regard.

Throughout Pakistani society in general, there is a growing weariness with the counterterrorism operations presently being waged on the country's soil. Recent polling, for example, suggests only weak support for using force against Islamic militants operating within Pakistan, and most respondents overwhelmingly oppose allowing outside forces to combat al-Qaeda on their national territory. A survey recently conducted by the Program on International Policy Attitudes at the University of Maryland in collaboration with the U.S. Institute of Peace found that just 44 percent of urban Pakistanis favored sending the Pakistan Army to the tribal areas to "pursue and capture al Qaeda fighters," and only 48 percent would allow the Pakistan Army to act against "Taliban insurgents who have crossed over from Afghanistan." In general, the survey concludes that "Pakistanis reject overwhelmingly the idea of permitting foreign troops to attack al-Qaeda on Pakistani territory. Four out of five (80 percent) say their government should not allow U.S. or other foreign troops to enter Pakistan to pursue and capture al Qaeda fighters," and three out of four (77 percent) oppose allowing foreign troops to attack Taliban insurgents based in Pakistan. Other polls reveal similar levels of disenchantment with the U.S.-supported campaign against terrorism. One report summarized it: ... despite their own concerns about terrorism, Pakistanis overwhelmingly oppose U.S.led efforts to fight terrorism— six- in- ten (59%) oppose America's anti- terror campaign, while only 13% back it. Like many other Muslim publics throughout Asia, the Middle East, and elsewhere, Pakistanis also oppose other key facets of U.S. foreign policy. Three- quarters (76%) say the U.S. should remove its troops from Iraq, and a similar proportion (75%) believe the U.S. and NATO should withdraw from Afghanistan, which shares a 1,500 mile border with Pakistan.

But Pakistanis are not just worried about the use of U.S. force in neighboring countries. They also fear they could become a target. More than seven- in- ten (72%) are very or somewhat worried that the U.S. could become a military threat to their country. And 64% name the U.S. as one of the countries posing the greatest potential threat to Pakistan, more than even long-standing arch- rival India (45%), with whom Pakistan has fought three major wars in the last sixty years.

Musharraf has attempted to cope with this increasing national weariness and to circumvent the problems caused by his army's operations, minimize its casualties, and soothe the roiling political environment in the tribal areas by episodic strategies of appeasement built around so- called peace accords with the pro-Taliban locals in South and North Waziristan. Under these accords, the indigenous residents were tasked with preventing cross- border movements of terrorists into Afghanistan and further attacks on

Pakistani civilian and military targets. They were also to ensure either the ejection or the surrender of all foreigners, meaning the non–South Asian cadres loyal to al- Qaeda, from the FATA in exchange for which the Pakistan Army would withdraw to its barracks, suspend its combat operations against the terrorists, and defer to the tribes in regard to resolving disputes relating to the status of particular individuals.

Musharraf's understandable objective in pursuing such a solution was to restore the status quo ante— hold the tribes responsible for maintaining peace and security as they had done traditionally— but it was a strategy that was doomed to failure because it did not appreciate the extent of radicalization in the FATA and the tribes' new determination to protect their al-Qaeda and Taliban cortege against the Pakistani government and the United States, which were viewed as the greater threats. Thus, although several tribal groups have sought to cooperate with the government in rooting out the radicals in their midst, the more extremist entities, not surprisingly, used the breathing space provided by the accords to recruit, train, and rearm the terrorists in anticipation of a heightened and continuing campaign in Afghanistan. The Pakistan Army's attack on the Lal Masjid in Islamabad proved to be the proverbial straw that finally destroyed the charade embodied by the peace accords in the FATA, but the failure of these agreements has left Musharraf in an unenviable limbo where neither peace nor war seems able to deliver the counterterrorism goals pursued by the Pakistani state.

Third, the operational context surrounding the counterterrorism effort in the tribal areas and in Afghanistan has changed considerably— to the disadvantage of the Western coalition— since Operation Enduring Freedom began in 2001. To begin with, the Taliban movement, which was never a tight and cohesive political entity in any case, has become an even looser network of affiliated individuals and groups since it was forced from power in Kabul. Today, the Taliban "alliance" can be characterized as a disparate congeries of several elements united only by a common religious ideology, a desire to regain power in either Afghanistan or their local areas of operation, and a deep antagonism toward the United States and its regional allies. Several distinct elements can be identified in the current Taliban coalition:

- The leadership shura centered around Mullah Omar and his cohort in Quetta and the subsidiary war councils in Quetta, Miran Shah, Peshawar, and Karachi;
- The Taliban cadres who survived the defeat in Afghanistan, which are loosely controlled by the regional shuras and continue to draw on the madaris in the FATA and the refugee camps in Pakistan for their continuing recruitment;
- The tribal networks of former mujahideen commanders like Jalaluddin Haqqani who operates in Paktika, Paktia, and Khowst provinces and provides a key bridge between al-Qaeda and the Taliban; Gulbuddin Hekmatyar who leads the Hezb-i-Islami and operates in Nangarhar, Konar, and Nuristan provinces; Anwar- ul- Haq who leads the Hezb-i-Islami (Khalis) also operates in the Nangarhar area and is believed to lead the Tora Bora Military Front; and Saifullah Mansoor, a veteran field commander who is known to be active in the eastern areas;
- The Pakistani Taliban commanders like Baitullah Mahsud, the chieftain of the Mahsud tribe in South Waziristan; Maulana Faqir Muhammad who is associated with the Tehrike- Nafaz- e- Shariat- e- Muhammad and who operates in the Bajaur Agency; Maulana Qazi Fazlullah, also affiliated with the same group but operating out of Swat; Mangal Bagh Afridi, who leads the Lashkar- e- Islami in the Khyber Agency and is believed to be part of a larger local opposition network led by Mufti Munir Shakir; and Sharif Khan and Nur Islam, tribal leaders who have demonstrated considerable operational effectiveness in South Waziristan;
- The drug lords in eastern and southern Afghanistan, especially in Helmand and Kandahar provinces, who are either taxed or willingly contribute revenues that are indispensable for the Taliban war against Kabul;
- The sundry former anti- Soviet commanders who control small groups of fighters and are engaged primarily in criminal activities such as bank robberies, kidnappings for

ransom, and assassination of local officials while they simultaneously offer their services as guns for hire;

- The disaffected Afghan Pashtun tribes, most conspicuously the rural Ghilzai, who, feeling disenfranchised in the current governing arrangements, continue to support the Taliban with manpower and sanctuary within Afghanistan; and, finally,
- Al- Qaeda, which, while distinct from all the foregoing groups in that its focus of operations remains the global jihad, nonetheless collaborates with the Taliban in order to assist the Taliban in recovering control of Kabul while it continues to preserve its sanctuary in the FATA in the interim.

The implication of such a diverse target set is that destroying the Taliban today has become much more difficult because its previously weak hierarchical structure has become even more diffuse, with truly diverse entities coordinating as necessary but with each also carrying out its own local agenda.

This reality, in turn, implies that while some specific nodes in this network will have to be defeated "kinetically" if the Taliban threat is to be erased, these tactical successes will have to be procured despite the political hesitation in parts of the Pakistani state and the real operational limitations of the Pakistani military. The complexity of Islamabad's relations with many of the constituent elements in the Taliban coalition does not help either: although Islamabad may readily cooperate in targeting some of the Pakistani Taliban commanders, drug lords, petty former anti-Soviet captains, and al- Qaeda elements, the ties nurtured by Pakistan's military and intelligence services with the Taliban leadership and the tribal networks of key former mujahideen commanders make these targets relatively inviolate, at least in the near term. For understandable sociopolitical reasons, Pakistani leaders are also likely to find it very difficult to conduct any large-scale interdiction operations aimed at the Taliban foot soldiers and the disaffected tribes— even if only in the FATA— partly because of the unmanageable chaos that would ensue in a very sensitive area of great importance to the Pakistani state and partly because of the fact that the insurgents drawn from these groups today are truly protean, capable of participating in military operations when required but at other times fluidly mutating into ordinary tribals.

There is no doubt, therefore, that winning the war on terror in Afghanistan will require dealing with the sanctuary enjoyed by various militant groups inside Pakistan. But it is probably an exaggeration to conclude "that the solution lies not in Afghanistan, but across the Khyber Pass in Pakistan."80 What happens in Afghanistan itself is critically important—not only in regard to ongoing military operations but, more fundamentally, in respect to reconstruction, economic development, nation building, and political reconstitution—because the counterterrorism campaign will not be won until the political environment in Afghanistan improves to the point where these insurgent forces are denied the conditions that allow them to survive and flourish. As General James L. Jones succinctly stated in his testimony before the Senate Foreign Relations Committee, "I am convinced that the solution in Afghanistan is not a military one."

It is in this context that Afghanistan's— and not just Pakistan's—political failures are particularly galling. Despite the advances in developing a constitutional government in Kabul with strong international support, the Karzai regime has turned out to be conspicuously ineffective. The inability of the government to deliver basic services, education, justice, and economic development, even in those areas not directly threatened by the insurgency, has fueled great frustration with President Karzai throughout the country. The growing corruption witnessed at all levels of government only exacerbates this resentment. And the runaway upsurge in poppy cultivation, which in 2006 yielded an all- time- high output of 6,100 metric tons, has resulted in a situation where "militia commanders, criminal organizations, and corrupt officials have exploited narcotics as a reliable source of revenue and patronage, which has perpetuated the threat these groups pose to the country's fragile internal security and the legitimacy of its embryonic democratic government." The complexities of intra- Afghan politics only compound the situation further: many Pashtun groups, for example, stung by the government's inability or unwillingness to address their specific grievances, often view the local insurgents as more effective instruments for achieving their immediate security or developmental goals. Any efforts made by the government to assuage Pashtun bitterness directly, however, complicates its relations with the non- Pashtun groups, who are apt to see most initiatives aimed at bolstering central authority, reinvigorating the traditional Pashtun tribal structures, and negotiating with Pakistan as evidence of a surreptitious attempt to reassert Pashtun hegemony over the rest of Afghanistan.

The Karzai government has thus far not succeeded in steering clear of these competing pressures, and its sharply alternating policies have not helped its standing either. Its most recent stab at neutralizing the growing insurgency by implementing a reconciliation program involving the "moderate Taliban" is a good example. After resisting such an idea for a long time, in part because of opposition from former Northern Alliance figures supportive of the government, President Karzai changed course and embarked on an effort to reintegrate the less extreme Taliban members into the national mainstream.

The idea is indeed sensible in principle but difficult to implement successfully in practice. That President Musharraf is its most ardent advocate has not raised the credibility of the program particularly, because it is often viewed in Afghanistan as a Pakistani stratagem to evade fulfilling its obligations to erase the insurgent sanctuaries in the tribal areas. In any event, the notion of reconciling moderate Taliban into Afghan society, while certainly commendable if it is understood to mean coopting the poor and disenfranchised confederations such as the Ghilzai, is invariably tricky and possibly even counterproductive because of the difficulties of distinguishing genuinely alienated individuals, who might be desirous of integration, from their more diehard and utterly intractable counterparts. One very thoughtful analysis concluded:

While more efforts should have gone into reconciliation in the early days, seeking to quell the insurgency now by rewarding criminal behavior would only perpetuate a culture of impunity and betray the trust of those who have backed the new, democratic, participatory institutions. It appears that the concept of reconciliation is being used interchangeably with amnesty. While such compromise may bring some measure of short- term relief, it would ultimately do nothing to break the cycle of violence.

Not surprisingly then, this Program Takhim- e Solh, which translates roughly into the Strengthening Peace program, has not been a noteworthy success. It does not appear to have made any significant dent in the manpower available to the insurgency, even as it has increased the fissures between Karzai and his non- Pashtun allies, has created new political threats to his 2009 presidential ambitions, and has failed to undermine the Taliban's social base of support in the interim.

One news report summarized the current crisis within Afghanistan laconically by declaring: "Government corruption and poppy cultivation are rampant and public services remain a wreck; food prices are soaring, unemployment remains high and resurgent Taliban forces in the south are pressing toward th[e] capital."86 Defeating this last threat obviously represents a classic chicken- and- egg dilemma: Taliban resurgence prevents the Karzai regime from effectively extending central control in the east, south, and southeast of the country, while the lack of effective state presence in these areas is precisely what makes the Taliban's return possible in the first place. Unfortunately, the three critical elements that could help Afghanistan break out of this cruel trap are constrained for different reasons.

To begin with, and as the foregoing discussion has elaborated, Pakistan is hobbled by political hesitancy and myriad operational limitations. To make things worse, NATO forces in Afghanistan are constrained by various "national caveats," that is, operational restrictions that prevent the alliance's International Security Assistance Force (ISAF) from undertaking the necessary combat operations required to prevent the Taliban from consolidating their foothold in southern and southeastern Afghanistan. Although ISAF is formally charged with the provision of security throughout Afghanistan, the main thrust of its effort revolves around supporting the approximately thirty- four provincial reconstruction teams (PRTs) operating throughout the country.

The PRTs are joint military- civilian units of between sixty and one thousand personnel engaged in reconstruction activities aimed at enhancing local security in order to permit the larger nation-building exercise to succeed. This essentially humanitarian mission, however, presumes the consent of local residents for its success and requires the PRT to be neutral and impartial with respect to any overarching political rivalries that may otherwise characterize its operating environment. The PRT is supposed to be robust enough to defend itself against attack, but it is emphatically not intended to be the spear point of change through offensive military action. Although the vision underlying the PRT is defensible, the fact that the social change it engenders has consequences for the local balance of power within Afghanistan implies that its activities are entirely unwelcome to those who oppose the larger nation-building project. Consequently, the PRTs specifically and the international community's reconstruction activities more generally have become the target of concerted attacks by Taliban and al-Qaeda forces, especially in southern and eastern Afghanistan. It is in this context that the national caveats handicapping ISAF— some seventy- one at last count— become relevant because they prevent the various contingents that compose the 38,000-strong force from effectively engaging the adversary as required in accordance with tactical necessity.

Thus, although ISAF's area of operations has now extended to all of Afghanistan thanks to its stage 3 expansion to the South and its stage 4 expansion to the East, the differential rules of engagement under which each national contingent now operates ensures that its full combat power cannot be brought to bear uniformly over the entire battlefield— to the obvious advantage of its Taliban and al- Qaeda adversaries. To make things worse, although ISAF is declared to be NATO's highest- priority mission, the alliance has had enormous difficulty convincing its member states to make the requisite contributions of manpower, equipment, and finance to secure victory in the combat operations in Afghanistan. As things stand today, NATO fields some 1.2 soldiers per thousand Afghan inhabitants. Even if the 85,000 Afghan security personnel and the 12,000-odd U.S. forces dedicated to Operation

Enduring Freedom are added to the number, the ratio of security forces to population hovers at about 4 soldiers per thousand inhabitants. This level of force presence is abysmal, given that a considerable body of research suggests that successful nation-building operations require at least 10 soldiers per thousand inhabitants, and preferably 20 soldiers per thousand inhabitants if there is an active conflict.

NATO thus far has simply not been able to contemplate, let alone provide, combat forces at anything approximating these levels required for success in Afghanistan. The British, the Canadians, and the Dutch have supplied the largest contingents actually involved in combat operations; although the Germans and the Italians have a significant presence, they are not involved in active combat. The French, too, while politically supportive of the ISAF mission in Afghanistan have declined to support the effort through either enhanced contributions or a restructuring of their current force posture. Although French forces remain among the most capable units within ISAF, the thousand- odd troops currently present in Afghanistan are deployed in the relatively secure areas in and around Kabul, with Paris continuing to resist NATO entreaties to dispatch these forces to the eastern and southern areas of Afghanistan where the Taliban opposition is most active.

The NATO presence in Afghanistan is thus characterized by a curious paradox: the most capable European states, largely those in western Europe, have simply declined to make the robust contributions required to win what is universally acknowledged as the "good war," while the alliance's newest entrants from eastern Europe appear far more willing to contribute to the ISAF effort even though they lack the depth of national and military resources possessed by their western European counterparts. Although the early U.S. disinclination to involve NATO in Afghan peace operations played some role in sustaining this paradox, the later European disenchantment with the U.S. war in Iraq, Washington's treatment of terrorist detainees, and the U.S. emphasis on attacking the adversary as opposed to protecting the population all appear to have contributed toward the western European inclination to stay aloof from any war- fighting entanglements.

This reluctance to contribute on the part of the stronger European allies is also reinforced partly by the presence of competing security priorities and the remarkably weak domestic support for any foreign military operations.

Above all else, however, it is owed to the alliance's failure to create a consensus on the implications of a failure in Afghanistan for European security; the lack of a clarifying continental debate on the goals, strategy, and tactics associated with winning the war against al- Qaeda and the Taliban; and the stunning unwillingness on the part of the wealthier "post- heroic" European states to actually fight a war that would require them to expend blood and treasure by remaining ensconced, deployed, and operating with their full panoply of military capabilities in southern and eastern Afghanistan until the adversary is eventually routed.

Finally, the principal combatant elements conducting Operation Enduring Freedom, through war- fighting actions against the al-Qaeda and Taliban forces in Afghanistan, are also handicapped by several limitations.

Among the most important of these is the availability of deployable troops. The 10,000-odd soldiers that the United States contributes to this combat operation, supplemented by token forces provided by the United Kingdom, Canada, and a few others, are increasingly insufficient given the growing scale and intensity of the Afghan insurgency.

Although coalition forces are superbly trained and equipped and have proved themselves devastatingly effective in combat with their Taliban adversaries, they are simply insufficient to maintain the large- scale presence that is now required to win the war in southern and eastern Afghanistan, given the virulence of the challenge.

In the initial phase of Operation Enduring Freedom, when military operations essentially consisted of search- and- destroy operations that targeted roving bands of terrorists, the current force size was probably appropriate because the superior mobility, firepower, and training of U.S. forces permitted them to sanitize large areas of territory despite their relatively small numbers. With the Taliban insurgency now nourished by local roots particularly in the Nimruz, Helmand, Kandahar, Oruzgan, and Zabol provinces, simply defeating the insurgents in battle is insufficient because they appear able to replenish their numbers relatively easily using both local recruits and imports from across the border, and, more important, they are able to return covertly to the contested districts after their tactically victorious U.S. adversaries withdraw to their rear bases.

Defeating this strategy requires a long- term presence of military forces in situ, which the division- sized U.S. combatant elements simply cannot provide. NATO's inability— and reluctance— to fulfill this role and to include combat operations whenever required merely compounds the problem with the result that the local inhabitants, especially those opposed to the Taliban, are compelled to make their peace with the insurgents merely as a means of preserving their security.

The surreptitious return of the Taliban in this fashion to any given district invariably results then in the ejection of the nongovernmental organizations working therein, the interruption of state- organized developmental assistance, and a crisis in local governance— all of which exacerbate the vicious circle that further undermines security. Given this dynamic, there is good reason to fear that, just as in Vietnam several decades earlier, U.S. and allied military forces could win every tactical engagement with the Taliban and yet lose the general war for Afghanistan.

Attempting to defeat this problem with the small number of combatant forces available only ends up overusing them. The Fort Drum, New York–based U.S. 10th Mountain Division, for example, has already deployed to Afghanistan thrice in five years, and when deployed in the field it is invariably overextended operationally because of the need to cover those alliance contingents that are unable to engage in combat operations when required because of their nationally constrained rules of engagement.

The only solution to this problem will be to beef up the U.S. military presence in southern Afghanistan: this may necessitate deploying more troops and equipment but, more importantly, entails creating a dedicated U.S. command operating under NATO aegis with full and exclusive authority to conduct the war as required without being hampered by the need to support the less able allied contingents.

The slow progress in raising the indigenous Afghan security forces has not helped either in the interim.

Most NATO officials interviewed on this subject declared flatly that the Afghan National Police is currently incapable of satisfactorily performing even basic law enforcement functions and, further, that it would be a long time before the Afghan National Army would be capable of operating as a tactical partner with allied units in counterinsurgency operations and even longer before it could do so independently.

Last, and despite many recent improvements, the coordination among the multiple national militaries engaged in reconstruction, counternarcotics, and counterterrorism in Afghanistan has simply not been as effective as it could be. In fact, problems of command and control among various coalition elements remain a serious and continuing impediment to the success of the Afghan stabilization effort. The vicious interaction of these many problems implies that the chicken- and- egg dilemmas confronting Afghanistan in regard to security and state presence are unlikely to be resolved any time soon.

On balance, therefore, the failure to eliminate the al- Qaeda and Taliban cadres in Afghanistan is owed to a complex cluster of causes. Pakistan's initial reluctance to interdict the Taliban stragglers who settled in the FATA and the leadership shuras that found homes in Quetta and elsewhere in Pakistan played an important role in permitting this organization to regenerate.

This process has been aided, however, by Afghanistan's own missteps in governance, including the failure to deliver security and economic and social development as well as to limit the runaway expansion in the cultivation of poppy in southern Afghanistan. The Pakistani effort to systematically interdict al-Qaeda while simultaneously going easy on the Taliban is riddled with inescapable contradictions. Although this strategy provided some early and important fruit, its tensions are now exploited by both the Taliban and al- Qaeda as well as by the vitally important tribal constituencies that are increasingly less bystanders than full participants on the wrong side in the ongoing war on terror in the FATA.

The operational, technical, and organizational limitations of the Pakistani counterterrorism forces deployed in some of the most inhospitable terrain in the world provides the final component of the explanation for why Islamabad has not done better. There is thus no doubt that Pakistan's reluctance to prosecute counterterrorism operations indivisibly has played an essential role in the failures along the Afghanistan- Pakistan border, but it is by no means the whole story, and any political posturing that suggests otherwise contributes neither to our understanding of the problem nor to its resolution.

9

Allegations of Support System in Pakistan for Terrorism

Allegations of a support system in Pakistan for Osama bin Laden have been made both before and after Osama bin Laden was found living in a compound in Abbottabad, Pakistan and was killed by a team of United States Navy SEALs on May 2, 2011. The United States had direct evidence that the ISI chief, Lt. Gen. Ahmed Shuja Pasha, knew of Bin Laden's presence in Abbottabad, Pakistan.

The compound itself was located just half a mile from Pakistan's premier military training academy in Abbottabad. In the aftermath of bin Laden's capture and death, U.S. President Barack Obama asked Pakistan to investigate the network that sustained bin Laden. "We think that there had to be some sort of support network for bin Laden inside of Pakistan," Obama said in a "60 Minutes" interview with CBS news. He also added that U.S. was not sure "who or what that support network was." In addition to this, in an interview to *Time* magazine, CIA Director Leon Panetta stated that U.S. officials did not alert Pakistani counterparts to the raid because they feared the terrorist leader would be warned.

According to Fred Burton, vice-president of the global intelligence firm Stratfor, officials of ISI, Pakistani military, along with one retired Pakistani military general, had knowledge of the arrangements made for bin Laden and safe house. Bin Laden's compound was razed that day at his Abbottabad safe house.

David Ignatius in *Washington Post* referred to the claim of the former ISI chief General Ziauddin Butt that the Abbottabad compound was used by Intelligence Bureau and noted that a report in the Pakistani press in December had quoted him as saying that Osama's stay at Abbottabad was arranged by Brigadier (retired) Ijaz Shah, senior ISI officer and the head of the Intelligence Bureau during 2004-2008, on Pervez Musharraf's orders. Later Butt denied making any such statement.

Allegations

"It was decided that any effort to work with the Pakistanis could jeopardize the mission. They might alert the targets." —CIA Chief Leon Panetta

Critics cited the very close proximity (800 yards) of bin Laden's heavily fortified compound (a custom-built luxury complex) to the Pakistan Military Academy, Pakistan's "West Point", and that the U.S. chose not to notify Pakistani authorities before the operation, and the alleged double standards of Pakistan regarding the perpetrators of the 2008 Mumbai attacks. U.S. government files leaked by Wikileaks disclosed that American diplomats were told that Pakistani security services were tipping off bin Laden every time U.S. forces approached. Pakistan's Inter-Services Intelligence (ISI) also helped smuggle al Qaeda militants into Afghanistan to fight NATO troops. According to the leaked files, in December 2009, the government of Tajikistan had told U.S. officials that many in Pakistan were aware of bin Laden's whereabouts.

The CIA Director and the US defense secretary Leon Panetta stated that Pakistan was "either involved or incompetent." In an interview to CBC television, 3 Obviously the concern has always been how could a compound like this, how could bin Laden be in an area where there were military establishments, where we could see the military operating and not have them know.

Regarding US's decision to withhold intelligence about the raid from Pakistan, he said, "The concern we had is that...we had provided intelligence to them with regards to other areas and unfortunately, for one way or another, it got leaked to the individuals we were trying to go after, so as a result of that we were concerned that if we were going to perform a sensitive mission like this, we had to do it on our own."

Talking about the support network Panetta stated that some "lower rank" officers in the military knew where Bin Laden was hiding. The defense secretary said, "Well, you know, these situations sometimes, the leadership within Pakistan [sic] is obviously not aware of certain things and yet people lower down in the military establishment find it very well, they`ve been aware of it,...But bottom line is that we have not had evidence that provides that direct link."

"It is inconceivable that bin Laden did not have support system in the country that allowed him to remain there for extended period of time," —Obama's chief counterterrorism advisor, John O. Brennan Husain Haqqani the former-Pakistan Ambassador to the United States, who earlier said both countries "cooperated in making sure" that the operation leading to bin Laden's death was "successful", has admitted that Osama bin laden indeed had a support system in Pakistan, albeit without the Pakistani government being privy to this fact.

"Obviously, bin Laden did have a support system (in Pakistan). The issue was that support system within the government and the state of Pakistan or within the society of Pakistan,...We all know that there are people in Pakistan who share the same belief system as bin Laden and other extremists... So that is a fact, that there are people who probably protected him... We did not know. We had no knowledge [about Bin Laden]. And if we had knowledge, we would have acted upon it long ago" —Mr. Haqqani said in an interview.

A Pakistani official, speaking anonymously, said "We assisted only in terms of authorization of the helicopter flights in our airspace" and that "we did not want anything to do with such an operation in case something went wrong."

U.S. Senator Joe Lieberman, Chairman of the Senate Homeland Security Committee, said "This is going to be a time of real pressure on Pakistan to basically prove to us that they didn't know that bin

Laden was there". John O. Brennan, Obama's chief counterterrorism advisor, said that it was inconceivable that bin Laden did not have support from within Pakistan. He further stated: "People have been referring to this as hiding in plain sight. We are looking at how he was able to hide out there for so long." Senator Dianne Feinstein said that "it's hard for me to understand how the Pakistanis... would not know what was going on inside the compound", and that top Pakistan officials may be "walking both sides of the street." Senator Lindsey Graham questioned, "How could [bin Laden] be in such a compound without being noticed?", raising suspicions that Pakistan was either uncommitted in the fight against Islamist militants or was actively sheltering them while pledging to fight them. A Pakistani intelligence official said that they had passed on raw phone tap data to U.S. that led to the operation, but had failed to analyse this data themselves. Carl Levin who is chairman United States Senate Committee on Armed Servicesstated that he believes Pakistani officials knew the location of bin Laden and had "no doubt" they also know the location of other senior al-Qaeda operatives. He said Pakistan's intelligence and army have "got a lot of explaining to do," given that bin Laden was holed up in such a large house with surrounding buildings, the fact that its residents took the unusual step of burning their garbage and avoiding any trash collection. He further stated, "It's hard to imagine that the military or police did not have any ideas what was going on inside of that." After the raid, the U.S. asked that Pakistan identify its top intelligence operatives as it tried to establish if any of them had contact with bin Laden in the last few years.

"Bin Laden was the 'Golden Goose' that the army had kept under its watch but which, to its chagrin, has now been stolen from under its nose. Until then, the thinking had been to trade in the Goose at the right time for the right price, either in the form of dollars or political concessions" —Prof. Pervez Hoodbhoy, Pakistani nuclear physicist, essayist and political-defence analyst, in *The Express Tribune* Mosharraf Zaidi, a leading Pakistani columnist, stated, "It seems deeply improbable that bin Laden could have been where he was killed without the knowledge of some parts of the Pakistani state."

WikiLeaks had revealed that a US diplomatic dispatch told the Americans that "many" inside Pakistan knew where bin Laden was. The document stated that "In Pakistan, Osama Bin Laden wasn't an invisible man, and many knew his whereabouts in North Waziristan, but whenever security forces attempted a raid on his hideouts, the enemy received warning of their approach from sources in the security forces."

Indian Minister for Home Affairs P. Chidambaram said that bin Laden hiding "deep inside" Pakistan was a matter of grave concern for India, and showed that "many of the perpetrators of the Mumbai terror attacks, including the controllers and the handlers of the terrorists who actually carried out the attack, continue to be sheltered in Pakistan". He called on Pakistan to arrest them.

The Globe and Mail reported local police saying that the compound belonged to Hizbul Mujahideen, a militant group supported by ISI which is fighting Indian forces in Kashmir.

In October 2011, former Pakistani Army Chief, General Ziauddin Butt has asserted that Osama bin Laden was kept in an Intelligence Bureau safe house in Abbottabad by the then Director-General of the Intelligence Bureau of Pakistan (2004–2008), Brigadier Ijaz Shah. According to him, this had occurred with the "full knowledge" of former army chief General Pervez Musharraf and possibly that of current Chief of Army Staff (COAS) General Ashfaq Pervez Kayani.

According to a police report dated Jan. 17, 2012 Osama Bin Laden's youngest wife Amal Ahmad Abdul Fatah claims that except for the eight or nine months just after 9/11, Laden skipped from home to home in Peshawar, Swat and Haripur, Pakistan before settling in Abbottabad for about the last six years of his life.

Pakistan's Denial

In a 2005 interview, the then-president Musharraf emphatically denied bin Laden was in Pakistan, stating "One thing is very sure, let me assure you, that we are not going to hide him for a rainy day and then release him to take advantage."

Pakistan's president Asif Ali Zardari denied that his country's security forces may have sheltered Osama bin Laden, and called any supposed support for bin Laden by the Pakistani government "baseless speculation." The Pakistani foreign office issued a statement that "categorically denies" media reports that Pakistan's leadership, "civil as well as military, had any prior knowledge of the U.S. operation against Osama bin Laden."

Pakistan's former president Pervez Musharraf denied that officials in his country were responsible, calling bin Laden's presence in Pakistan a "blunder". Musharraf said instead that there was a possibility that rogue lower-level members of Pakistan's intelligence and military may have had knowledge of bin Laden's location. He conceded they might have known during the last year of his presidency six years ago, and said there ought to be an investigation.

"It's really appalling that he was there and nobody knew. I'm certainly appalled that I didn't know and that intelligence people from that time onward didn't know for 6 years that he was inside. And there is no excuse for this great, massive slip-up. And an investigation is in order and people must be punished for this big lapse."

"As a policy, the army and the ISI fighting terrorism and extremism, al Qaeda, Taliban. But rogue element within is a possibility," he said.

The head of Pakistan's powerful Directorate of Inter-Services Intelligence (ISI) Lieut. Gen. Ahmed Shuja Pasha, the ISI chief, conceded that Osama bin Laden's presence in Pakistan had been an "intelligence failure" but denied the ISI could have had any role in hiding Bin Laden. "If we had shielded Osama bin Laden, why would we have killed and arrested so many al-Qaeda leaders?" he stated "Would we have hidden such a large target in such an exposed area? Without any guards or escape route?"

FOLLOW-UP BY PAKISTANI AUTHORITIES

Regardless of Pakistan's claims about their previous conduct, many outside observers have raised ongoing concerns that most of the people jailed by them since Bin Laden's killing have been those who were trying to help capture him, rather than those who helped shield him.

One such incident would be when Dr. Shakil Afridi who assisted the CIA in the hunt for Osama bin Laden, was arrested several weeks after the killing of bin Laden. A Pakistani court imposed a 33-year sentence Wednesday on the doctor. The doctor was eventually tried under a tribal judicial system that denies the accused the right to have an attorney or to present evidence. According to *Washington post*, the Doctor could have received the death penalty if he had been tried under normal Pakistani law.

3 It is now indisputable that militancy in Pakistan is supported by the ISI (Inter-Services Intelligence). Pakistan's fight against militancy is bogus. It's just to extract money from America, Dr. Shakil Afridi stated in an interview with Fox News. He was interviewed from inside Peshawar Central Jail, where he is being held by Pakistani authorities. According to his statements, the Pakistani authorities said The Americans are our worst enemies, worse than the Indians and he suffered 2 crippling torture and 2 psychological abuse during the 12 months he was held by Pakistani spy agency for helping the US. He further revealed the fact that the ISI is in collusion with terrorists.

Abbottabad Commission Report

A judge-led inquiry set up by Pakistani government in 2011, based on interviews with 201 sources found there was evidence of incompetence at every level in the Pakistan's intelligence and security services and it did not rule out the involvement of rogue elements within the Pakistani intelligence service. The 336-page Abbottabad Commission Report, obtained in July 2013 by Al Jazeera, blasted Pakistan's civilian and military leadership for "gross incompetence" over the bin Laden affair. It found that by 2005, Pakistani intelligence was no longer actively pursuing intelligence that could lead to his capture. The report called the handling of the bin Laden situation a "natural disaster" and even called on the leadership to apologize to the people of Pakistan for their "dereliction of duty." Al Jazeera reported that the government's intention in conducting the inquiry was likely aimed at "regime continuance, when the regime is desperate to distance itself from any responsibility for the national disaster that occurred on its watch" and was likely "a reluctant response to an overwhelming public and parliamentary demand." Lack of intelligence on bid Laden's nine-year residence in the country was blamed on "Government Implosion Syndrome." Lack of knowledge of a CIA support network without Pakistan being aware was "a case of collective and sustained dereliction of duty by the political, military, and intelligence leadership." Although the report focused on the night of the raid, it had other findings. One was that bid Laden had been living in Pakistan since 2002, after surviving the Battle of Tora Bora. Another was that he and some family members moved into the compound in Abbottabad in 2005, the same year Pakistani intelligence stopped independently looking for him.

10

Al-Qaeda and Spiritual Terrorism

AL-QAEDA

Al-Qaeda, alternatively spelled al-Qaida and sometimes al-Qa'ida, is an international Sunni Islamist movement founded in 1988. Al-Qaeda have attacked civilian and military targets in various countries, the most notable being the September 11 attacks in 2001. These actions were followed by the US government launching a military and intelligence campaign against al-Qaeda called the *War on Terror*.

Characteristic techniques include suicide attacks and simultaneous bombings of different targets. Activities ascribed to it may involve members of the organization, who have taken a pledge of loyalty to Osama bin Laden, or the much more numerous "al-Qaeda-linked" individuals who have undergone training in one of its camps in Afghanistan or Sudan but not taken any pledge. Al-Qaeda's objectives include the end of foreign influence in Muslim countries and the creation of a new Islamic caliphate. Reported beliefs include that a Christian-Jewish alliance is conspiring to destroy Islam, and that the killing of bystanders and civilians is Islamically justified in jihad. Its management philosophy has been described as "centralization of decision and decentralization of execution." Following 9/11 and the launching of the War on Terrorism, it is thought al-Qaeda's leadership has "become geographically isolated", leading to the "emergence of decentralized leadership" of regional groups using the al-Qaeda "brand name."

Al-Qaeda has been labeled a *terrorist organization* by the United Nations Security Council, the North Atlantic Treaty Organization Secretary General, the Commission of the European Communities of the European Union, the United States Department of State, the Australian Government, Government of India, Public Safety Canada, the Israel Ministry of Foreign Affairs, Japan's Diplomatic Bluebook, South Korean Foreign Ministry, the Dutch Military Intelligence and Security Service, the United Kingdom Home Office, Russia, the Swedish Ministry for Foreign Affairs, and the Swiss Government.

ETYMOLOGY

In Arabic, *al-Qaeda* has four syllables. However, since two of the Arabic consonants in the name the voiceless uvular plosive [q] and the voiced pharyngeal fricative are not phones found in the English language. Al-Qaeda's name can also be transliterated as al-Qaida, al-Qa'ida, el-Qaida, or al Qaeda. The name of the organization comes from the Arabic noun *qâ'idah,* which means *foundation* or *basis* and can also refer to a military base or database. The initial *al-*is the Arabic definite article *the,* hence *the base*. In Arabic qa'idah bayanat is database where bayanat is data and qa'idah is base.

Osama bin Laden explained the origin of the term in a videotaped interview with Al Jazeera journalist Tayseer Alouni in October 2001:

> *"The name 'al-Qaeda' was established a long time ago by mere chance. The late Abu Ebeida El-Banashiri established the training camps for our mujahedeen against Russia's terrorism. We used to call the training camp al-Qaeda. The name stayed. "*

Saad Al-Faqih, a Saudi expert on al-Qaeda, has stated that the name al-Qaeda, "...originated from a documentation system in the Bait al-Ansar guesthouse back in the 1980s." The United Kingdom politician Robin Cook, who served as the United Kingdom Foreign Secretary and Leader of the House of Commons described Al-Qaeda as meaning "the database" and a product of western miscalculation. Cook wrote, "Al-Qaeda, literally "the database", was originally the computer file of the thousands of mujahideen who were recruited and trained with help from the CIA to defeat the Russians."

HISTORY OF THE NAME

There is at least one public reference to the name "al-Qaeda" that pre-dates the 2001 trial. The name appears with the spelling "al-Qaeda" in an executive order issued by President Bill Clinton in 1998, less than two weeks after the bombings of the U.S. embassies in Kenya and Tanzania. Executive Order 13099, issued on August 20, 1998, lists the organization as one of several associated with Osama bin Laden, the others being the Islamic Army, Islamic Salvation Foundation, the Islamic Army for the Liberation of the Holy Places, The World Islamic Front for Jihad Against Jews and Crusaders, and The Group for the Preservation of the Holy Sites.

The name *al-Qaeda* could have been introduced to the U.S. intelligence community by Jamal al-Fadl, who had been providing the Central Intelligence Agency with intelligence about bin Laden since 1996. The defecting Al-Fadl was debriefed by the CIA's Bin Laden Station ("Alex Base"). In the 1998 United States embassy bombings, al-Fadl testified that al-Qaeda was established in either late 1989 or early 1990 to continue the jihad after the Soviet withdrawal from Afghanistan. He said during the war against the Soviets, bin Laden had been funding a group called Maktab al-Khadamat, which was led by Abdallah Azzam. This organization was based in Pakistan and provided training, money and other support for Muslims who would cross the border into Afghanistan to fight.

According to al-Fadl, the Maktab al-Khadamat was disbanded following the Soviet withdrawal, but bin Laden wanted to establish a new group to continue the jihadist cause on other fronts. Al-Fadl testified that al-Qaeda's leader was initially Abu Ayoub al-Iraqi, who was later replaced by Abu Ubaidah al-Banshiri, but that both of these leaders nevertheless "reported to" bin-Laden. Al-Fadl claims the group initially went by two different names "al-Qaeda" and "Islamic Army", before eventually settling on the former. A meeting was apparently held in Khost, Afghanistan to establish the new group, which al-Fadl claims to have attended. Al-Fadl's recollection was that this occurred in either late 1989 or early 1990. Journalist Peter Bergen argues that two documents seized from the Sarajevo office of the Benevolence International Foundation show that the organization was established in August 1988. Both of these documents contain minutes of meetings held to establish a new military group and contain the term "al-qaeda".

Author Lawrence Wright also quotes this document (an exhibit from the "Tareek Osama" document presented in *United States v. Enaam M. Arnaout*), in his book *The Looming Tower.* Notes of a meeting of bin Laden and others on August 20, 1988 indicate "the military base" (*"al-qaeda al-askariya"*), was a formal group: `basically an organized Islamic faction, its goal is to lift the word of God, to make His religion victorious.` A list of requirements for membership itemized "listening and obedient... good manners" and making a pledge (bayat) to obey superiors. According to Wright, "[t]he name al-Qaeda was not used," in public pronouncements like the 1998 fatwa to kill Americans and their allies because "its existence was still a closely held secret." Wright writes that Al-Qaeda was formed at a August 11, 1988 meeting of "with several senior leaders" of Egyptian Islamic Jihad, (Sayyed Imam Al-Sharif, Ayman Al-Zawahiri, and others), Abdullah Azzam, and Osama bin Laden, where it was agreed to join bin Laden's money with the expertise of the Islamic Jihad organization and continue jihad elsewhere after the Soviets withdrew from Afghanistan.

In April 2002, the group assumed the name *Qa'idat al-Jihad,* which means "the base of Jihad". According to Diaa Rashwan, this was "...apparently as a result of the merger of the overseas branch of Egypt's al-Jihad (EIJ) group, led by Ayman El-Zawahiri, with the groups Bin Laden brought under his control after his return to Afghanistan in the mid-1990s."

HISTORY

Jihad in Afghanistan

The origins of the group can be traced to the Soviet war in Afghanistan. The United States viewed the conflict in Afghanistan, with the Afghan Marxists and allied Soviet troops on one side and the native Afghan mujahedeen on the other, as a blatant case of Soviet expansionism and aggression. The U.S. channelled funds through Pakistan's Inter-Services Intelligence agency to the native Afghan mujahedeen fighting the Soviet occupation in a CIA program called Operation Cyclone.

At the same time, a growing number of foreign Arab mujahedeen (also called Afghan Arabs) joined the jihad against the Afghan Marxist regime, facilitated by international Muslim organizations, particularly the Maktab al-Khidamat, whose funds came from some of the $600 million a year donated to the jihad by the Saudi Arabia government and individual Muslims- particularly wealthy Saudis who were approached by Osama bin Laden. Maktab al-Khidamat was established by Abdullah Azzam and Bin Laden in Peshawar, Pakistan, in 1984. From 1986 it began to set up a network of recruiting offices in the United States, the hub of which was the Al Kifah Refugee Center at the Farouq Mosque in Brooklyn's Atlantic Avenue. Among notable figures at the Brooklyn center were "double agent" Ali Mohamed, whom FBI special agent Jack Cloonan called "bin Laden's first trainer," and "Blind Sheikh" Omar Abdel-Rahman, a leading recruiter of mujahideen for Afghanistan.

The Afghan Mujahedeen of the 1980s have been alleged to be the inspiration for terrorist groups in nations such as Indonesia, the Philippines, Egypt, Saudi Arabia, Algeria, Chechnya, and the former Yugoslavia. According to Russian sources, the perpetrators of the first World Trade Center bombing in 1993 allegedly used a manual allegedly written by the CIA for the Mujihadeen fighters in Afghanistan on how to make explosives.

Al-Qaeda evolved from the Maktab al-Khidamat (Services Office), a Muslim organization founded in 1980 to raise and channel funds and recruit foreign mujahadeen for the war against the Soviets in Afghanistan. It was founded by Abdullah Yusuf Azzam, a Palestinian Islamic scholar and member of the Muslim Brotherhood. Maktab al-Khadamat organized guest houses in Peshawar, in Pakistan, near the Afghan border, and paramilitary training camps in Afghanistan to prepare international non-Afghan recruits for the Afghan war front. Azzam persuaded Bin Laden to join MAK, to use his own money and use his connections with "the Saudi royal family and the petro-billionaires of the Gulf" to raise more to help the mujahideen. The role played by MAK and foreign Muslim volunteers, or "Afghan Arabs", in the war was not a major one. While 250,000 Afghan Mujahideen fought the Soviets and the communist Afghan government, it is estimated that were never more than 2000 foreign mujahideen in the field at any one time. Nonetheless, foreign mujahedeen volunteers came from 43 countries and the number that participated in the Afghan movement between 1982 and 1992 is reported to have been 35,000.

The Soviet Union finally withdrew from Afghanistan in 1989. To the surprise of many, Mohammed Najibullah's communist Afghan government hung on for three more years before being overrun by elements of the mujahedeen. With mujahedeen leaders unable to agree on a structure for governance, chaos ensued, with constantly reorganizing alliances fighting for control of ill-defined territories, leaving the country devastated.

Expanding Operations

Toward the end of the Soviet military mission in Afghanistan, some mujahedeen wanted to expand their operations to include Islamist struggles in other parts of the world, such as Israel and Kashmir. A number of overlapping and interrelated organizations were formed to further those aspirations.

One of these was the organization that would eventually be called al-Qaeda, formed by Osama bin Laden with an initial meeting held on August 11, 1988. Bin Laden wished to establish nonmilitary operations in other parts of the world; Azzam, in contrast, wanted to remain focused on military campaigns. After Azzam was assassinated in 1989, the MAK split, with a significant number joining bin Laden's organization.

In November 1989, Ali Mohamed, a former special forces Sergeant stationed at Fort Bragg, North Carolina, left military service and moved to Santa Clara, California. He traveled to Afghanistan and Pakistan and became "deeply involved with bin Laden's plans.". A year later, on November 8, 1990, the FBI raided the New Jersey home of Mohammed's associate El Sayyid Nosair, discovering a great deal of evidence of terrorist plots, including plans to blow up New York City skyscrapers. Nosair was eventually convicted in connection to the 1993 World Trade Center bombing, and for the murder of Rabbi Meir Kahane on November 5, 1990. In 1991, Ali Mohammed is said to have helped orchestrate Osama bin Laden's relocation to Sudan.

Gulf War and the Start of U.S. Enmity

Following the Soviet Union's withdrawal from Afghanistan, Osama bin Laden returned to Saudi Arabia. The Iraqi invasion of Kuwait in 1990 had put the country of Saudi Arabia and its ruling House of Saud at risk as Saudi's most valuable oil fields (Hama) were within easy striking distance of Iraqi forces in Kuwait, and Saddam's call to pan-Arab/Islamism could potentially rally internal dissent. In the face of a seemingly massive Iraqi military presence, Saudi Arabia's own forces were well armed but far outnumbered. Bin Laden offered the services of his mujahedeen to King Fahd to protect Saudi Arabia from the Iraqi army. The Saudi monarch refused bin Laden's offer, opting instead to allow U.S. and allied forces to deploy on Saudi territory.

The deployment angered Bin Laden, as he believed the presence of foreign troops in the "land of the two mosques" (Mecca and Medina) profaned sacred soil. After speaking publicly against the Saudi government for harboring American troops, he was quickly forced into exile to Sudan and on April 9, 1994 his Saudi citizenship was revoked. His family publicly disowned him. There is controversy over whether and to what extent he continued to garner support from members of his family and/or the Saudi government.

Sudan

From approximately 1992 to 1996, al-Qaeda and bin Laden were located in Sudan, coming at the invitation of Islamist theoretician Hassan al Turabi following an Islamist coup d'état, and leaving after being expelled by the Sudanese government. During this time bin Laden assisted the Sudanese government, bought or set up various business enterprises, and established training camps where insurgents trained. But in Sudan bin Laden lost his Saudi passport and source of income in response to his verbal attacks on the Saudi king. A key turning point for bin Laden occurred in 1993 when Saudi Arabia gave support for the Oslo Accords which set a path for peace between Israel and Palestine.

Zawahiri and the EIJ, who served as the core of al-Qaeda but also engaged in separate operations against the Egyptian government, had even worse luck in Sudan. In 1993, a young schoolgirl was killed in an unsuccessful EIJ attempt on the life of the Egyptian Interior Minister, Hasan al-Alfi. Egyptian public opinion turned against Islamist bombings and the police arrested 280 more of al-Jihad's members and executed six. In 1995 an even more ill-fated attempt to assassinate Egyptian president Mubarak led to the expulsion of EIJ and not long after of bin Laden by the Sudanese government.

Refuge in Afghanistan

After the Soviet withdrawal, Afghanistan was effectively ungoverned for seven years and plagued by constant infighting between former allies and various mujahedeen groups.

Throughout the 1990s, a new force began to emerge. The origins of the Taliban (literally "students") lay in the children of Afghanistan, many of them orphaned by the war, and many of whom had been educated in the rapidly expanding network of Islamic schools (madrassas) either in Kandahar or in the refugee camps on the Afghan-Pakistani border.

According to Ahmed Rashid, five leaders of the Taliban were graduates of a single madrassa, Darul Uloom Haqqania (also known as "the University of Jihad",) in the small town of Akora Khattak near Peshawar, situated in Pakistan but largely attended by Afghan refugees. This institution reflected Salafi beliefs in its teachings, and much of its funding came from private donations from wealthy Arabs, for whom bin Laden provided conduit. A further four leading figures (including the perceived Taliban leader Mullah Mohammed Omar Mujahed) attended a similarly funded and influenced madrassa in Kandahar, Afghanistan. Many of the mujahedeen who later joined the Taliban fought alongside Afghan warlord Mohammad Nabi Mohammadi's Harkat i Inqilabi group at the time of the Russian invasion. This group also enjoyed the loyalty of most Afghan Arab fighters.

The continuing internecine strife between various factions, and accompanying lawlessness following the Soviet withdrawal, enabled the growing and well-disciplined Taliban to expand their control over territory in Afghanistan, and they came to establish an enclave which it called the Islamic Emirate of Afghanistan. In 1994, they captured the regional center of Kandahar, and after making rapid territorial gains thereafter, conquered the capital city Kabul in September 1996.

After Sudan made it clear that bin Laden and his group were no longer welcome that year, Taliban-controlled Afghanistan—with previously established connections between the groups, a similar outlook on world affairs and largely isolated from American political influence and military power—provided a perfect location for al-Qaeda to establish its headquarters. Al-Qaeda enjoyed the Taliban's protection and a measure of legitimacy as part of their Ministry of Defense, although only Pakistan, Saudi Arabia, and the United Arab Emirates recognized the Taliban as the legitimate government of Afghanistan.

Al-Qaeda training camps in Afghanistan and the Pakistani border regions are alleged to have trained militant Muslims from around the world. Despite the perception of some people, al-Qaeda members are ethnically diverse and connected by their radical version of Islam.

An ever-expanding network of supporters thus enjoyed a safe haven in Taliban-controlled Afghanistan until the Taliban were defeated by a combination of local forces and United States air power in 2001 (see section *September 11, attacks and the United States response*). Osama bin Laden and other al-Qaeda leaders are still believed to be located in areas where the population is sympathetic to the Taliban in Afghanistan or the border Tribal Areas of Pakistan.

Fatwas

In 1996, al-Qaeda announced its jihad to expel foreign troops and interests from what they felt were Islamic lands. Bin Laden issued a fatwa, which amounted to a public declaration of war against the United States and any of its allies, and began to focus al-Qaeda's resources towards attacking the United States and its interests. Also occurring on June 25, 1996 was the bombing of the Khobar towers, located in Khobar, Saudi Arabia. On February 23, 1998, Osama bin Laden and Ayman al-Zawahiri, a leader of Egyptian Islamic Jihad, along with three other Islamist leaders, co-signed and issued a fatwa (binding religious edict) under the banner of the World Islamic Front for Combat Against the Jews and Crusaders (*al-Jabhah al-Islamiyya al-'Alamiyya li-Qital al-Yahud wal-Salibiyyin*) declaring:

[T]he ruling to kill the Americans and their allies-civilians and military — is an individual duty for every Muslim who can do it in any country in which it is possible to do it, in order to liberate the al-Aqsa Mosque (in Jerusalem) and the holy mosque (in Makka) from their grip, and in order for their armies to move out of all the lands of Islam, defeated and unable to threaten any Muslim. This is in accordance with the words of Almighty Allah, 'and fight the pagans all together as they fight you all together,' and 'fight them until there is no more tumult or oppression, and there prevail justice and faith in Allah'.

Neither bin Laden nor al-Zawahiri possessed the traditional Islamic scholarly qualifications to issue a fatwa of any kind; however, they rejected the authority of the contemporary ulema (seen as the paid servants of *jahiliyya* rulers) and took it upon themselves. Assassinated former FSB agent Alexander Litvinenko alleged that the Russian FSB trained al-Zawahiri in a camp in Dagestan eight months before the 1998 fatwa.

ORGANIZATION STRUCTURE

Though the current structure of al-Qaeda is unknown, information mostly acquired from Jamal al-Fadl provided American authorities with a rough picture of how the group was organized. While the veracity of the information provided by al-Fadl and the motivation for his cooperation are both disputed, American authorities base much of their current knowledge of al-Qaeda on his testimony.

Osama bin Laden is the emir and Senior Operations Chief of al-Qaeda (although originally this role may have been filled by Abu Ayoub al-Iraqi), advised by a Shura Council, which consists of senior al-Qaeda members, estimated by Western officials at about twenty to thirty people. Ayman al-Zawahiri is al-Qaeda's Deputy Operations Chief and Abu Ayyub al-Masri is possibly the senior leader of al-Qaeda in Iraq.

- The Military Committee is responsible for training operatives, acquiring weapons, and planning attacks.
- The Money/Business Committee runs business operations, provides air tickets and false passports, pays al-Qaeda members, and oversees profit-driven businesses. In the 9/ 11 Commission Report, it is estimated that al-Qaeda requires $30,000,000 USD per year to conduct its operations.
- The Law Committee reviews Islamic law and decides if particular courses of action conform to the law.
- The Islamic Study/Fatwah Committee issues religious edicts, such as an edict in 1998 telling Muslims to kill Americans.
- In the late 1990s there was a publicly known Media Committee, which ran the now-defunct newspaper *Nashrat al Akhbar (Newscast)* and handled public relations.
- In 2005, al Qaeda formed As-Sahab, a media production house, to supply its video and audio materials.

The number of individuals belonging to the organization is also unknown. According to the controversial BBC documentary *The Power of Nightmares*, al-Qaeda is so weakly linked together that it is hard to say it exists apart from Osama bin Laden and a small clique of close associates. The lack of any significant numbers of convicted al-Qaeda members despite a large number of arrests on terrorism charges is cited by the documentary as a reason to doubt whether a widespread entity that meets the description of al-Qaeda exists at all. Therefore the extent and nature of al-Qaeda remains a topic of dispute.

Its rank and file has been described as changing from being "predominantly Arab," in its first years of operation, to "largely

Pakistani," as of 2007. It has been estimated that 62% of al-Qaeda members have university education.

Organization v. Concept

When asked about the possibility of Al Qaeda's connection to the 7 July 2005 London bombings in 2005, Metropolitan Police Commissioner Sir Ian Blair said:

> *"Al Qaeda is not an organization. Al Qaeda is a way of working... but this has the hallmark of that approach.... Al Qaeda clearly has the ability to provide training... to provide expertise... and I think that is what has occurred here."*

What exactly al-Qaeda is, or was, remains in dispute. In the BBC documentary *The Power of Nightmares,* writer and journalist Adam Curtis contends that the idea of al-Qaeda as a formal organization is primarily an American invention. Curtis contends the name "al-Qaeda" was first brought to the attention of the public in the 2001 trial of Osama bin Laden and the four men accused of the 1998 United States embassy bombings in East Africa. As a matter of law, the U.S. Department of Justice needed to show that Osama bin Laden was the leader of a criminal organization in order to charge him *in absentia* under the Racketeer Influenced and Corrupt Organizations Act, also known as the RICO statutes. The name of the organization and details of its structure were provided in the testimony of Jamal al-Fadl, who claimed to be a founding member of the organization and a former employee of Osama bin Laden. To quote the documentary directly:

The reality was that bin Laden and Ayman Zawahiri had become the focus of a loose association of disillusioned Islamist militants who were attracted by the new strategy. But there was no organization. These were militants who mostly planned their own operations and looked to bin Laden for funding and assistance. He was not their commander. There is also no evidence that bin Laden used the term "al-Qaeda" to refer to the name of a group until after September the 11th, when he realized that this was the term the Americans had given it.

Questions about the reliability of al-Fadl's testimony have been raised by a number of sources because of his history of dishonesty and because he was delivering it as part of a plea bargain agreement after being convicted of conspiring to attack U.S. military establishments. Sam Schmidt, a defense lawyer from the trial, had the following to say about al-Fadl's testimony:

There were selective portions of al-Fadl's testimony that I believe was false, to help support the picture that he helped the Americans join together. I think he lied in a number of specific testimony about a unified image of what this organization was. It made al-Qaeda the new Mafia or the new Communists. It made them identifiable as a group and therefore made it easier to prosecute any person associated with al-Qaeda for any acts or statements made by bin Laden.

IDEOLOGY

The radical Islamist movement in general and al-Qaeda in particular developed during the Islamic revival and Islamist movement of the last three decades of the 20th century along with less extreme movements.

Some have argued that "without the writings" of Islamic author and thinker Sayyid Qutb "al-Qaeda would not have existed." Qutb preached that because of the lack of sharia law the Muslim world was no longer Muslim, having reverted to pre-Islamic ignorance known as jahiliyyah. To restore Islam, a vanguard movement of righteous Muslims was needed to implement Sharia and rid the Muslim world of any non-Muslim influences, such as concepts like socialism or nationalism. Enemies of Islam included "treacherous Orientalists" and "world Jewry", who plotted "conspiracies" and "wicked[ly]" opposed Islam.

In the words of Mohammed Jamal Khalia, a close college friend of Osama bin Laden: *Islam is different from any other religion; it's a way of life. We [Khalia and bin Laden] were trying to understand what Islam has to say about how we eat, who we marry, how we talk. We read Sayyid Qutb. He was the one who most affected our generation.*

Qutb had an even greater influence on Osama bin Laden's mentor and another leading member of al-Qaeda, Ayman al-Zawahiri. Zawahiri's uncle and maternal family patriarch, Mafouz Azzam, was Qutb's student, then protégé, then personal lawyer and finally executor of his estate-one of the last people to see Qutb before his execution.

"Young Ayman al-Zawahiri heard again and again from his beloved uncle Mahfouz about the purity of Qutb's character and the torment he had endured in prison." Zawahiri paid homage to Qutb in his work *Knights under the Prophet's Banner.*

One of the most powerful effects of Qutb's ideas was the idea that many who said they were Muslims were not, i.e. they were apostates, which not only gave jihadists "a legal loophole around the prohibition of killing another Muslim," but made "it a religious obligation to execute" the self-professed Muslim. These alleged apostates included leaders of Muslims countries since they failed to enforce sharia law.

ATTACKS

1992

On December 29, 1992, al-Qaeda's first terrorist attack took place as two bombs were detonated in Aden, Yemen. The first target was the Movenpick Hotel and the second was the parking lot of the Goldmohur Hotel. The bombings were an attempt to eliminate American soldiers on their way to Somalia to take part in the international famine relief effort, Operation Restore Hope. Internally, al-Qaeda considered the bombing a victory that frightened the Americans away, but in the United States the attack was barely noticed. No Americans were killed because the soldiers were staying in a different hotel altogether, and they went on to Somalia as scheduled. However little noticed, the attack was pivotal as it was the beginning of al-Qaeda's change in direction, from fighting armies to killing civilians. Two people were killed in the bombing, an Australian tourist and a Yemeni hotel worker. Seven others, mostly Yemenis, were severely injured.

Two *fatwa* are said to have been appointed by the most theologically knowledgeable of al-Qaeda's members, Mamdouh Mahmud Salim, aka Abu Hajer al Iraqi, to justify the killings according to Islamic law. Mamdouh Mahmud Salim referred to the thirteenth-century scholar Ibn Taymiyyah, much admired by

Wahhabis. In a famous fatwa, Ibn Tamiyyah had ruled that Muslims should kill the invading Mongols, and so too Salim said al-Qaeda should kill American soldiers. The second fatwa followed another of Ibn Tamiyyah's, that Muslims should not only kill Mongols but anyone who aided the Mongols, who bought goods from them or sold to them. In addition the killing of someone merely standing near a Mongol was justified as well. He ruled these killings just because any innocent bystander, like the Yemenite hotel worker, would find their proper reward in death, going to Paradise if they were good Muslims and to hell if they were bad. This became al-Qaeda's justification for killing civilians.

1993 WORLD TRADE CENTER BOMBING

In 1993, Ramzi Yousef used a truck bomb to attack the World Trade Center in New York City. The attack was intended to break the foundation of Tower One knocking it into Tower Two, bringing the entire complex down. Yousef hoped this would kill 250,000 people. The towers shook and swayed but the foundation held and he succeeded in killing only six people (although he injured 1,042 others and caused nearly $300 million in property damage).

After the attack, Yousef fled to Pakistan and later moved to Manila. There he began developing the Bojinka Plot plans to blow up a dozen American airliners simultaneously, to assassinate Pope John Paul II and President Bill Clinton, and to crash a private plane into CIA headquarters. He was later captured in Pakistan.

None of the U.S. government's indictments against Osama bin Laden have suggested that he had any connection with this bombing, but Ramzi Yousef is known to have attended a terrorist training camp in Afghanistan. After his capture, Yousef declared that his primary justification for the attack was to punish the United States for its support for the Israeli occupation of Palestinian territories and made no mention of any religious motivations.

Late 1990s

On November 13, 1995, a van containing a hundred pounds of Semtex explosive blew up near the communications center for the Saudi National Guard in downtown Riyadh, Saudi Arabia, where some American military contractors and Army officers had been training the Saudi National Guard. Seven people were killed, and sixty people were injured. The Saudi government arrested four men, "torturing confessions" out of them that they had been inspired by bin Laden's speeches and trained at al-Qaeda's camp in Afghanistan, and quickly executed them. It is unclear if they had anything to do with the crime. As with many bombings suspected to be the work of al-Qaeda, bin Laden praised the attacks but denied authorizing the attack or training the bombers.

The U.S. embassy bombings in East Africa, resulting in upward of 300 deaths, mostly locals. A barrage of cruise missiles launched by the U.S. military in response devastated an al-Qaeda base in Khost, Afghanistan, but the network's capacity was unharmed.

Bin Laden then turned his sights towards the United States Navy. In October 2000, al-Qaeda militants in Yemen bombed the missile destroyer U.S.S. Cole in a suicide attack, killing 17 U.S. servicemen and damaging the vessel while it lay offshore. Inspired by the success of such a brazen attack, al-Qaeda's command core began to prepare for an attack on the United States itself.

September 11 Attacks

The September 11 attacks were the most devastating terrorist acts in American and world history, killing approximately 3,000 people. Two commercial airliners were deliberately flown into the World Trade Center towers, a third into The Pentagon, a fourth, originally intended to target the United States Capitol crashed in Pennsylvania.

The attacks were conducted by al-Qaeda, acting in accord with the 1998 *fatwa* issued against the United States and its allies by military forces under the command of bin Laden, al-Zawahiri, and others. Evidence points to suicide squads led by al-Qaeda military commander Mohamed Atta as the culprits of the attacks, with bin Laden, Ayman al-Zawahiri, Khalid Shaikh Mohammed, and Hambali as the key planners and part of the political and military command. Messages issued by bin Laden after September 11, 2001 praised the attacks, and explained their motivation while denying any involvement.

Bin Laden legitimized the attacks by identifying grievances felt by both mainstream and Islamist Muslims, such as the general perception that the United States was actively oppressing Muslims. Bin Laden asserted that America was massacring Muslims in 'Palestine, Chechnya, Kashmir and Iraq' and that Muslims should retain the 'right to attack in reprisal'. He also claimed the 9/11 attacks were not targeted at women and children, but 'America's icons of military and economic power'.

Evidence has since come to light that the original targets for the attack may have been nuclear power stations on the east coast of the U.S. The targets were later altered by al-Qaeda, as it was feared that such an attack "might get out of hand".

War on Terrorism

In the immediate aftermath of the attacks, the United States government decided to respond militarily, and began to prepare its armed forces to overthrow the Taliban regime it believed was harboring al-Qaeda. Before the United States attacked, it offered Taliban leader Mullah Omar a chance to surrender bin Laden and his top associates. The Taliban offered to turn over bin Laden to a neutral country for trial if the United States would provide evidence of bin Laden's complicity in the attacks. U.S. President George W. Bush responded by saying: "We know he's guilty. Turn him over", and British Prime Minister Tony Blair warned the Taliban regime: "Surrender bin Laden, or surrender power". Soon thereafter the United States and its allies invaded Afghanistan, and together with the Afghan Northern Alliance removed the Taliban government in the war in Afghanistan.

As a result of the United States using its special forces and providing air support for the Northern Alliance ground forces, both Taliban and al-Qaeda training camps were destroyed, and much of the operating structure of al-Qaeda is believed to have been disrupted. After being driven from their key positions in the Tora Bora area of Afghanistan, many al-Qaeda fighters tried to regroup in the rugged Gardez region of the nation. Again, under the cover of intense aerial bombardment, U.S. infantry and local Afghan forces attacked, shattering the al-Qaeda position and killing or capturing many of the militants. By early 2002, al-Qaeda had been dealt a serious blow to its operational capacity, and the Afghan invasion appeared an initial success. Nevertheless, a significant Taliban insurgency remains in Afghanistan, and al-Qaeda's top two leaders, bin Laden and al-Zawahiri, evaded capture.

Debate raged about the exact nature of al-Qaeda's role in the 9/11 attacks, and after the U.S. invasion began, the U.S. State Department also released a videotape showing bin Laden speaking with a small group of associates somewhere in Afghanistan shortly before the Taliban was removed from power. Although its authenticity has been questioned by some, the tape appears to implicate bin Laden and al-Qaeda in the September 11 attacks and was aired on many television channels all over the world, with an accompanying English translation provided by the United States Defense Department. In September 2004, the U.S. government commission investigating the September 11 attacks officially concluded that the attacks were conceived and implemented by al-Qaeda operatives.

In October 2004, bin Laden appeared to claim responsibility for the attacks in a videotape released through Al Jazeera, saying he was inspired by Israeli attacks on high-rises in the 1982 invasion of Lebanon: "As I looked at those demolished towers in Lebanon, it entered my mind that we should punish the oppressor in kind and that we should destroy towers in America in order that they taste some of what we tasted and so that they be deterred from killing our women and children."

By the end of 2004, the U.S. government claimed that two-thirds of the top leaders of al-Qaeda from 2001 were in custody (including Ramzi bin al-Shibh, Khalid Sheikh Mohammed, Abu Zubaydah, Saif al Islam el Masry, and Abd al-Rahim al-Nashiri) or dead (including Mohammed Atef). Despite the capture or death of many senior al-Qaeda operatives, the U.S. government continues to warn that the organization is not yet defeated, and battles between U.S. forces and al-Qaeda-related groups continue. In the meantime, autonomous regional branches of al-Qaeda continue to emerge around the world.

REGIONAL ACTIVITIES

Africa

Al-Qaeda involvement in Africa has included a number of bombing attacks in North Africa, as well as supporting parties in civil wars in Eritrea and Somalia. From 1991 to 1996, Osama bin Laden and other Al-Qaeda leaders were based in the Sudan.

Asia

Al-Qaeda involvement in Asia has largely been centered on Afghanistan and northwest sections of Pakistan. Al-Qaeda also has affiliate groups, including Abu Sayyaf in the Philippines, and has carried out a number of attacks in Indonesia. For a period of time in the mid-1990s, Khalid Sheikh Mohammed and Ramzi Yousef resided in Manila in the Philippines, and were plotting attacks including Oplan Bojinka.

Europe

European activities of Al Qaeda have included involvement in the wars in Bosnia, as well as attacks carried out in Istanbul, London, and Madrid.

Middle East

In the Middle East, Al-Qaeda has been involved in a number of attacks inside Saudi Arabia, occurring as early as 1995. Al Qaeda or Ayman al-Zawahiri's former organization, Egyptian Islamic Jihad, have carried out various attacks inside Egypt. In Iraq, elements at first loosely associated with al-Qaeda, in the Jama'at al-Tawhid wal-Jihad organization commanded by Abu Musab al-Zarqawi, have played a key role in the War in Iraq.

INTERNET ACTIVITIES

In the wake of its evacuation from Afghanistan, al-Qaeda and its successors have migrated online to escape detection in an atmosphere of increased international vigilance. As a result, the organization's use of the Internet has grown more sophisticated, encompassing financing, recruitment, networking, mobilization, publicity, as well as information dissemination, gathering, and sharing.

Abu Ayyub al-Masri's al-Qaeda movement in Iraq regularly releases short videos glorifying the activity of jihadist suicide bombers. In addition, both before and after the death of Abu Musab al-Zarqawi (the former leader of al-Qaeda in Iraq), the umbrella organization to which al-Qaeda in Iraq belongs, the Mujahideen Shura Council, has a regular presence on the Web where pronouncements are given by Murasel. This growing range of multimedia content includes guerrilla training clips, stills of victims about to be murdered, testimonials of suicide bombers, and epic-themed videos with high production values that romanticize participation in jihad through stylized portraits of mosques and musical scores. A website associated with al-Qaeda, for example, posted a video of captured American entrepreneur Nick Berg being decapitated in Iraq. Other decapitation videos and pictures, including those of Paul Johnson, Kim Sun-il, and Daniel Pearl, were first posted on jihadist websites. With the rise of "locally rooted, globally inspired" terrorists, counter-terrorism experts are currently studying how al-Qaeda is using the Internet–through websites, chat rooms, discussion forums, instant messaging, and so on – to inspire a worldwide network of support.

The bombers responsible for the 7 July 2005 London bombings, some of whom were well integrated into their local communities, are an example of such "globally inspired" terrorists, and they reportedly used the Internet to plan and coordinate. A group called the "Secret Organization of al-Qaeda in Europe" claimed responsibility for the London bombings on the Al-Qalah website, which was subsequently shut down. The publicity opportunities offered by the Internet have been particularly exploited by al-Qaeda. In December 2004, for example, bin Laden released an audio message by posting it directly to a website, rather than sending a copy to al Jazeera as he had done in the past. Al Qaeda turned to the Internet for release of its videos in order to be certain it would be available unedited, rather than risk the possibility of al Jazeera editors editing the videos and cutting out anything critical of the Saudi royal family. Bin Laden's December 2004 message was much more vehement than usual in this speech, lasting over an hour.

In the past, Alneda.com and Jehad.net were perhaps the most significant al-Qaeda websites. Alneda was initially taken down by American Jon Messner, but the operators resisted by shifting the site to various servers and strategically shifting content. The U.S. is currently attempting to extradite an information technology specialist, Babar Ahmad, from the UK, who is the creator of various English-language al-Qaeda websites such as Azzam.com. Ahmad's extradition is opposed by various British Muslim organizations, such as the Muslim Association of Britain.

ALLEGED CIA INVOLVEMENT

Whether or not the al-Qaeda attacks were " blowback" from the American CIA's "Operation Cyclone" program to help the Afghan mujahideen is a matter of some debate. Robin Cook, former British Foreign Secretary from 1997-2001, has written that al-Qaeda and Bin Laden were "a product of a monumental miscalculation by western security agencies," and that the mujahideen that formed al-Qaeda were "originally... recruited and trained with help from the CIA".

A variety of sources — CNN journalist Peter Bergen, Pakistani ISI Brigadier Mohammad Yousaf, and CIA operatives involved in the Afghan program, such as Vincent Cannistraro and Milton Bearden — deny that the CIA or other American officials had contact with the Afghan Arabs (foreign mujahideen) or Bin Laden, let alone armed, trained, coached or indoctrinated them. Bergen and others argue that there was no need to recruit foreigners unfamiliar with the local language, customs or lay of the land since there were a quarter of a million local Afghans willing to fight; that Arab Afghans themselves had no need for American funds since they received several hundred million dollars a year from non-American, Muslim sources; that Americans could not have trained mujahideen because Pakistani officials would not allow more than a handful of them to operate in Pakistan and none in Afghanistan; and that the Afghan Arabs were almost invariably militant Islamists reflexively hostile to Westerners whether or not the Westerners were helping the Muslim Afghans.

According to Peter Bergen, known for conducting the first television interview with Osama bin Laden in 1997, the idea that "the CIA funded bin Laden or trained bin Laden... a folk myth. There's no evidence of this.... Bin Laden had his own money, he was anti-American and he was operating secretly and independently.... The real story here is the CIA didn't really have a clue about who this guy was until 1996 when they set up a unit to really start tracking him." But as Bergen himself admitted, in one "strange incident" the CIA *did* appear to give visa help to mujahideen-recruiter Omar Abdel-Rahman.

CRITICISM

According to a number of sources there has been a "rising tide of anger in the Islamic world toward Al Qaeda and its affiliates" by "religious scholars, former fighters, and militants... alarmed" by Al Qaeda's takfir and killing of Muslims in Muslim countries, especially Iraq. Noman Benotman, a former Afghan Arab and militant of the Libyan Islamic Fighting Group, went public with an open letter of criticism to Ayman al-Zawahiri in November 2007 after persuading imprisoned senior leadership of his former group to enter into peace negotiations with the Libyan regime. While Ayman al-Zawahiri announced the affiliation of the group with Al Qaeda in November 2007, the Libyan government released 90 members of the group from prison several months later after "they were said to have renounced violence."

In 2007, around the sixth anniversary of September 11, Sheikh Salman al-Ouda, a Saudi religious scholar and one of the fathers of the Sahwa, the fundamentalist awakening movement that swept through Saudi Arabia in the '80s, delivering a personal rebuke to Osama bin Laden. Al Ouda addressed Al Qaeda's leader on MBC, a widely watched Middle East TV network, asking him

My brother Osama, how much blood has been spilt? How many innocent people, children, elderly, and women have been killed... in the name of Al Qaeda? Will you be happy to meet God Almighty carrying the burden of these hundreds of thousands or millions [of victims] on your back? In 2007, the imprisoned Sayyed Imam Al-Sharif, an influential Afghan Arab, "ideological godfather of Al Qaeda", and former supporter of takfir, sensationally withdrew his support from al Qaeda with a book *Wathiqat Tarshid Al-'Aml Al-Jihadi fi Misr w'Al-'Alam* ("Document of Right Guidance for Jihad Activity in Egypt and the World").

Usama Hassan, an Imam in London and former supporter of Al Qaeda who traveled to Afghanistan to train as jihadi was alienated by the July 2005 bombings in London and now preaches against bin Laden and helped launch the Quilliam Foundation.

According to Pew polls, support for Al Qaeda has been dropping around the Muslim world in recent years. The numbers supporting suicide bombings in Indonesia, Lebanon, and Bangladesh, for instance, have dropped by half or more in the last five years. In Saudi Arabia, only 10 percent now have a favorable view of Al Qaeda, according to a December poll by Terror Free Tomorrow, a Washington-based think tank.

IMPACT OF AFGHAN WAR ON PAKISTAN

Afghanistan the neighboring country of Pakistan has been in war since one can remember. From Mongol Emperors to Russian invasion and now the country is facing war on terror. These wars have shattered the base of Afghanistan and left it in ruins. It lacks proper administration, governance, and communication infrastructure.

The Statistics available are not accurate as they should be. The estimated population of Afghanistan is about 31, 900,000 from which 28% are literate.

Overall View

Where Afghanistan has suffered from wars that has happened from time to time its neighboring countries have also suffered including Pakistan and Iran. After Russian invasion of Afghanistan in 1979, more than 2.8 million Afghan refugees sought shelter in Pakistan and about 1.5 million have fled to Iran.

Russia lost almost 15,000 soldiers and many wounded. In 1988 the United States, Pakistan, Afghanistan, and the Soviet Union signed an agreement by which the Russia would withdraw its troops, which completed in 1989. After Russian invasion a group of tribal and urban groups arose and all of these known collectively as the mujahidin and afterward they form into now known Taliban. Even after the end of Russian invasion, the Afghan refugees have stayed in Pakistan and caused extra burden on the economy of our country and they prove as cheap labour causing unemployment to local workers.

Most of the Afghan immigrants have no documentation of any kind. Hardly any of them have ID or any record to their permanent location.

This has given some of culprit's an edge because they are not afraid to be identified or recognized. Most of the drug dealers, illegal arms dealer, car lifters, or smugglers are Afghan refugees that have made their permanent shops in the borders areas.

The easy access to Illegal weaponry has promoted the Kalakov Culture and has made weaponry very easily available even to common person at a very reasonable price.

For example, the 9mm pistol can be easily available at the low price of only Rs 3600, US$ 45 that is made in Dara Adam Khail. A point worth mentioning is that the weapons are made legal by puncturing, the original weapon caliber number with new number and then District Coordination Officer sanctions the weapon.

Drugs Smuggling and Effects on Trade

The Taliban's main source of income has been drug dealing and production. They gathered their resources from Opium cultivation according to data from CSIS (Central for Strategic and International Studies) Opium cultivation has continued to grow but has shifted into Taliban controlled areas.

An opium survey was conducted by UN in 2007 which showed production of 185 Metric Ton opium in 2001 and 8200 Metric Tons in 2007.

There are more than 20 unofficial border crossing points on Pak Afghan border from where opium is transferred to Pakistan in NWFP and Baluchistan Province. The trade route between Afghanistan and Pakistan has also been affected. Trade Route Torkham Jalalabad highway now considered as a danger zone.

Dry fruits that were imported to Pakistan, have now reached sky high prices such as nuts, Walnuts, Almonds etc.

War on Terror

After 9/11 America invaded Afghanistan to end the Al-Qaeda and Taliban Regime but things seem to have gotten worse for both countries. Pakistan is now under pressure by US leadership to wipe out the Terrorist camps in the Northern borders of Pakistan. Pakistan has played a front line allay in the war against terror since 2001. According to a report released by an US Department in 2004, Pakistan had faced a loss of 10 billion by becoming an ally on war on terror.

Assuming that in 2004 we were not in as much problems as now then after three years as a front line ally we have lost 40 billion minimum. Taliban had only one rival group Northern Alliance in Afghanistan. NA was supported by India and Russia to fight against Taliban.

When US attacked Afghanistan, Taliban was in great number and their only rival group Northern Alliance leaded by Ahmad Shah Massoud was US main hope. Eventually US supported NA in fight against Taliban. With time Taliban regime began to lose hold but did not end completely.

The US army for the time became confident of their victory but their dreams shattered when Taliban rose again with more strength. By the end of 2006 there were more than 123 suicide attacks in Afghanistan and 69 in Pakistan. In assessment report of Pakistan by Institute for Conflict Management Database at least 3,599 persons, including 1,523 civilians, 597 security force (SF) personnel and 1,479 militants, were killed in 2007. These suicide attacks began after the US attack on Afghanistan.

Political Impacts

Pakistan had supported Taliban in war against Russia. Russia and India supported Northern Alliance Ahmad Shah but after US invasion against Taliban and using NA to their advantage Pakistan got cornered and had to leave his old allies Taliban. The Taliban consists of people belonging to a cast known as Pachtons.

Most Pachtons live in the Pak Afghan Border area. US army used our Air Fields against Taliban such as Dera Ismail and Shahbaz Airbase which angered the Pachtons. Pak Army was also utilized in operations against the bases of Taliban in Northern, southern reigns and FATA. These things with time brought hatred in people of Northern and Southern regions that had a majority of Pachtons. The war on terror in Afghanistan has left us alone in many ways. In an survey of Afghan public opinion it shows that the Afghanis remains supportive of the US and major NATO/ISAF countries but the results showed high level hostility to Pakistan.

We had to change our border policies Iran and China were also annoyed when we allowed US to use our land their home security was also threatened by US. We became the welcoming guest of US in Asia. This has caused us isolation from all sides we are in a state of being bankrupt and US has used us and left us as always.

Borders

Another dilemma is that our Borders have become unsecured from all sides. Our army which was being used to protect Kashmiri people is now also utilized on Durand line.

Kashmir issue is becoming an issue of past and long forgotten even the dotted lines of Jammu Kashmir region is now removed from international maps and that region is shown as part of India. Kashmiri Mujahidin's were engaging Indian army and kept them occupied but due to our incorrect policies the Kashmir movement is becoming weak day by day.

This allowed Indian agencies to interfere in our conflicted regions supporting elements against Pakistan. Pak Intelligence has found evidence of outside meddling in FATA, Northern and Southern Regions.

Advance Tactical equipment of foreign built, imported military hardware, digital maps, and many other gadgets which can be only available to army were gathered from taken over militant camps. Other than that we now have to defend from day to day suicidal attacks. Almost 80,000 of our army are being utilized on Pak Afghan border.

Ethnic and Tribal Disputes

Growing pressure by US to increase military advancement is also becoming a hurdle rather than progress. The Pashtun's and Baluchi living in FATA, NWFP, and Baluchistan are not something to overlook upon.

Pashtun are living in border areas of Afghanistan and Pakistan there are about 42% Pashtun in Afghanistan and 10% percent in Northern areas of Pakistan. Pashtun have blood relations in Afghanistan, after US attack, they became violent. Rage gathered in their hearts and minds that turned into clash between Pak army and Tribe leaders. We are now standing at a very sensitive point. The recent bombings and gun battles by the Pak army have forced 200,000 people to flee from their homes in Northern Ares.

Now and then women and children get killed. The tribal leaders want economic help to eradicate illiteracy and extremism but they have forgotten that economic giants start these wars. The thing we really want is to check our policies whether this war is right? The answer is easy. Wars are fought from slavery to freedom not from freedom to slavery.

TERRORISM EXPERTS' VIEW ON THE ALLIANCE

According to experts, Zarqawi gave al-Qaeda a highly visible presence in Iraq at a time when its original leaders went into hiding or were killed after the September 11, 2001, attacks in the United States. In turn, al-Qaeda leaders were able to brand a new franchise in Iraq and claim they were at the forefront of the fight to expel U.S. forces. But this relationship was proven to be fragile as Zarqawi angered al-Qaeda leaders by focusing attackings on Iraqi Shia's more often than U.S. military. In September 2005, U.S. intelligence officials said they had confiscated a long letter that al-Qaeda's deputy leader, Ayman al-Zawahiri, had written to Zarqawi, bluntly warning that Muslim public opinion was turning against him. According to Paul Wilkinson, chairman of the Center for the Study of Terrorism and Political Violence at the University of St. Andrews in Scotland, "A number of al-Qaeda figures were uncomfortable with the tactics he was using in Iraq...It was quite clear with Zarqawi that as far as the al-Qaeda core leadership goes, they couldn't control the way in which their network affiliates operated."

U.S. Officials' view on the Alliance

In June 2004, Secretary of Defense Donald Rumsfeld conceded that Zarqawi's ties to Al Qaeda may have been much more ambiguous—and that he may have been more a rival than a lieutenant to bin Laden. Zarqawi "may very well not have sworn allegiance to [bin Laden]," Rumsfeld said at a Pentagon briefing. "Maybe he disagrees with him on something, maybe because he wants to be 'The Man' himself and maybe for a reason that's not known to me." Rumsfeld added that, "someone could legitimately say he's not Al Qaeda."

According to the Senate Report on Prewar Intelligence released in September 2006, "in April 2003 the CIA learned from a senior al-Qa'ida detainee that al-Zarqawi had rebuffed several efforts by bin Ladin to recruit him. The detainee claimed that al-Zarqawi had religious differences with bin Ladin and disagreed with bin Ladin's singular focus against the United States. The CIA assessed in April 2003 that al-Zarqawi planned and directed independent terrorist operations without al Qaeda direction, but assessed that he 'most likely contracts out his network's services to al Qaeda in return for material and financial assistance from key al Qaeda facilitators.'"(page 90)

In the April 2006 National Intelligence Estimate, declassified in September 2006, it asserts that "Al-Qa'ida, now merged with Abu Mus'ab al-Zarqawi's network, is exploiting the situation in Iraq to attract new recruits and donors and to maintain its leadership role."

Alleged Links to Saddam Hussein

On February 5, 2003, then Secretary of State Colin Powell addressed the U.N. Security Council on the issue of Iraq. Regarding Zarqawi, Powell stated that:

> *"Iraq today harbors a deadly terrorist network headed by Abu Musab Al-Zarqawi, an associated in collaborator of Osama bin Laden and his Al Qaeda lieutenants. When our coalition ousted*

the Taliban, the Zarqawi network helped establish another poison and explosive training center camp. And this camp is located in northeastern Iraq. He traveled to Baghdad in May 2002 for medical treatment, staying in the capital of Iraq for two months while he recuperated to fight another day. During this stay, nearly two dozen extremists converged on Baghdad and established a base of operations there. These Al Qaeda affiliates, based in Baghdad, now coordinate the movement of people, money and supplies into and throughout Iraq for his network, and they've now been operating freely in the capital for more than eight months. We asked a friendly security service to approach Baghdad about extraditing Zarqawi and providing information about him and his close associates. This service contacted Iraqi officials twice, and we passed details that should have made it easy to find Zarqawi. The network remains in Baghdad."

Abu Musab al Zarqawi recuperated in Baghdad after being wounded while fighting along with Taliban and al-Qaeda fighters in Afghanistan. According to the 2004 Senate Report of Pre-war Intelligence on Iraq, "A foreign government service asserted that the IIS (Iraqi Intelligence Service) knew where al-Zarqawi was located despite Baghdad's claims that it could not find him."page 337 The Senate Report on Prewar Intelligence also stated "As indicated in Iraqi Support for Terrorism, the Iraqi regime was, at a minimum, aware of al-Zarqawi's presence in Baghdad in 2002 because a foreign government service passed information regarding his whereabouts to Iraqi authorities in June 2002. Despite Iraq's pervasive security apparatus and its receipt of detailed information about al-Zarqawi's possible location, however, Iraqi Intelligence told the foreign government service it could not locate al-Zarqawi."

Jordanian Analysis

A Jordanian security official told the *Washington Post* that documents recovered after the overthrow of Saddam show that Iraqi agents detained some of Zarqawi's operatives but released them after questioning. He also told the *Washington Post* that the Iraqis warned the Zarqawi operatives that the Jordanians knew where they were. The official also told the *Washington Post* that "'We sent many memos to Iraq during this time, asking them to identify his position, where he was, how he got weapons, how he smuggled them across the border,' but Hussein's government never responded."

This claim was reiterated by Jordanian King Abdullah II in an interview with *Al-Hayat*. Abdullah revealed that Saddam Hussein had rejected repeated requests from Jordan to hand over al-Zarqawi. According to Abdullah, "We had information that he entered Iraq from a neighboring country, where he lived and what he was doing. We informed the Iraqi authorities about all this detailed information we had, but they didn't respond." King Abdullah told the *Al-Hayat* that Jordan exerted "big efforts" with Saddam's government to extradite al-Zarqawi, but added that "our demands that the former regime hand him over were in vain.

One high-level Jordanian intelligence official told the *Atlantic Monthly* that al-Zarqawi, after leaving Afghanistan in December 2001, frequently traveled to the Sunni Triangle of Iraq where he expanded his network, recruited and trained new fighters, and set up bases, safe houses, and military training camps. He said, however, "We know Zarqawi better than he knows himself. And I can assure you that he never had any links to Saddam."

Counterterrorism scholar Loretta Napoleoni quotes former Jordanian parliamentarian Layth Shubaylat, who was personally acquainted with both Zarqawi and Saddam Hussein:

> *"First of all, I don't think the two ideologies go together, I'm sure the former Iraqi leadership saw no interest in contacting al-Zarqawi or al-Qaeda operatives. The mentality of al-Qaeda simply doesn't go with the Ba'athist one. When he was in prison in Jordan with Shubaylat, Abu Mos'ab wouldn't accept me, said Shubaylat, because I'm opposition, even if I'm a Muslim. How could he accept Saddam Hussein, a secular dictator?"*

U.S. Conclusion

A CIA report in late 2004 concluded that there was no evidence Saddam's government was involved or even aware of this medical treatment, and found no conclusive evidence the Saddam Hussein regime had harbored Zarqawi. A US official told *Reuters* that the report was a mix of new information and a look at some older information and did not make any final judgments or come to any definitive conclusions. "To suggest the case is closed on this would not be correct," the official said." A US official familiar with the report told Knight-Ridder that "what is indisputable is that Zarqawi was operating out of Baghdad and was involved in a lot of bad activities." Another U.S. official summarized the report as such: "The evidence is that Saddam never gave Zarqawi anything."

According to the 2004 Senate Report on Prewar Intelligence, "The CIA provided four reports detailing the debriefings of Abu Zubaydah, a captured senior coordinator for al-Qaeda responsible for training and recruiting. Abu Zubaydah said that he was not aware of a relationship between Iraq and al-Qaeda. He also said, however, that any relationship would be highly compartmented and went on to name al-Qaeda members who he thought had good contacts with the Iraqis. For instance, Abu Zubaydah indicated that he had heard that an important al-Qaeda associate, Abu Mus'ab al-Zarqawi, and others had good relationships with Iraqi Intelligence."

A classified memo obtained by Stephen F. Hayes, prepared by Undersecretary of Defense for Policy Douglas J. Feith in response to questions posed by the Senate Intelligence Committee as part of its investigation into prewar intelligence, stated the following regarding al-Zarqawi:

> *"Sensitive reporting indicates senior terrorist planner and close al Qaeda associate al Zarqawi has had an operational alliance with Iraqi officials. As of October 2002, al Zarqawi maintained contacts with the IIS to procure weapons and explosives, including surface-to-air missiles from an IIS officer in Baghdad. According to sensitive reporting, al Zarqawi was setting up sleeper cells in Baghdad to be activated in case of a U.S. occupation of the city, suggesting his operational cooperation with the Iraqis may have deepened in recent months. Such cooperation could include IIS provision of a secure operating bases [sic] and steady access to arms and explosives in preparation for a possible U.S. invasion. Al Zarqawi's procurements from the Iraqis also could support al Qaeda operations against the U.S. or its allies elsewhere."*

The memo was a collection of raw intelligence reports and drew no conclusions. U.S. intelligence officials conveyed to Newsweek that the "reports [in the memo] were old, uncorroborated and came from sources of unknown if not dubious credibility."

The 2006 Senate Report on Prewar Intelligence concluded that Zarqawi was not a link between Saddam and al-Qaeda: "Postwar information indicates that Saddam Hussein attempted, unsuccessfully, to locate and capture al-Zarqawi and that the regime did not have a relationship with, harbor, or turn a blind eye toward Zarqawi." The report also cited the debriefing of a "high-ranking Iraqi official" by the FBI. The official stated that a foreign government requested in October 2002 that the IIS locate five individuals suspected of involvement in the murder of Laurence Foley, which lead to the arrest of Abu Yasim Sayyem in early 2003. The official told the FBI that evidence of Sayyem's ties to Zarqawi was compelling, and thus, he was "shocked" when Sayemm was ordered released by Saddam. The official stated it "was ludicrous to think that the IIS had any involvement with al-Qaeda or Zarqawi," and suggested Saddam let Sayyem go because he "would participate in striking U.S. forces when they entered Iraq." In 2005, according to the Senate report, the CIA amended its 2004 report to conclude that "the regime did not have a relationship, harbor, or turn a blind eye toward Zarqawi and his associates."page 91–92 An intelligence official familiar with the CIA assessment also told Michael Isikoff of *Newsweek* magazine that the current draft of the report says that while Zarqawi did likely receive medical treatment in Baghdad in 2002, the report concludes that "most evidence suggests Saddam Hussein did not provide Zarqawi safe haven before the war,...[but] it also recognizes that there are still unanswered questions and gaps in knowledge about the relationship."

The Army's Foreign Military Studies Office website translated a letter dated August 17, 2002 from an Iraqi intelligence official. The letter is part of the Operation Iraqi Freedom documents. The letter asks agents in the country to be on the lookout for Abu Musab al-Zarqawi and another unnamed man. Pictures of both men were attached.

The letter issued the following 3 directives:

1. Instructing your sources to continue their surveillance of the above mentioned individuals in your area of operations and inform us once you initiate such action.
2. Coordinate with Directorate 18 to verify the photographs of the above mentioned with photos of the members of the Jordanian community within your area of operations.
3. Conduct a comprehensive survey of all tourist facilities (hotels, furnished apartments, and leased homes). Give this matter your utmost attention. Keep us informed.

The documents also contain responses to this request. One response, dated August 2002, states "Upon verifying the information through our sources and friends in the field as well as office, we found no information to confirm the presence of the above mentioned in our area of operation. Please review, we suggest circulating the contents of this message." Another response, also dated August 2002, states "After closely examining the data and through our sources and friends in (SATTS: U R A) square, and in Al-Qa'im immigration office, and in Office, none of the mentioned individuals are documented to be present in our area of jurisdiction."

According to *ABC news*, "The letter seems to be coming from or going to Trebil, a town on the Iraqi-Jordanian border. Follow up on the presence of those subjects is ordered, as well as a comparison of their pictures with those of Jordanian subjects living in Iraq. In his book *At the Center of the Storm*, George Tenet writes:

> *"...by the spring and summer of 2002, more than a dozen al-Qa'ida-affiliated extremists converged on Baghdad, with apparently no harassment on the part of the Iraqi government. They found a comfortable and secure envirnonment in which they moved people and supplies to support Zarqawi's operations in northern Iraq."*

According to Tenet, while Zarqawi did find a safe haven in Iraq and did supervise camps in northeastern Iraq run by Ansar al-Islam, "the intelligence did not show any Iraqi authority, direction, or control over any of the many specific terrorist acts carried out by al-Qa'ida."

Arguments Downplaying Zarqawi's Importance

Some people have claimed that Zarqawi's notoriety was the product of U.S. war propaganda designed to promote the image of a demonic enemy figure to help justify continued U.S. military operations in Iraq, perhaps with the tacit support of jihadi elements who wished to use him as a propaganda tool or as a distraction. In one report, the conservative newspaper *Daily Telegraph* described the claim that Zarqawi was the head of the "terrorist network" in Iraq as a "myth." This report cited an unnamed U.S. military intelligence source to the effect that the Zarqawi leadership "myth" was initially caused by faulty intelligence, but was later accepted because it suited U.S. government political goals. One Sunni insurgent leader claimed on 11 December that "Zarqawi is an American, Israeli and Iranian agent who is trying to keep our country unstable so that the Sunnis will keep facing occupation."

On February 18, 2006, Shiite cleric Muqtada al-Sadr made similar charges:

> *"I believe he is fictitious. He is a knife or a pistol in the hands of the occupier. I believe that all three-the occupation, the takfir (i.e. the practice of declaring other Muslims to be heretics) supporters, and the Saddam supporters-stem from the same source, because the takfir supporters and the Saddam supporters are a weapon in the hands of America and it pins its crimes on them."*

On April 10, 2006, the *Washington Post* reported that the U.S. military conducted a major propaganda offensive designed to exaggerate Zarqawi's role in the Iraqi insurgency. Gen. Mark Kimmitt says of the propaganda campaign that there "was no attempt to manipulate the press." In an internal briefing, Kimmitt is quoted as stating, "The Zarqawi PSYOP Program is the most successful information campaign to date." The main goal of the propaganda campaign seems to have been to exacerbate a rift between insurgent forces in Iraq, but intelligence experts worried that it had actually enhanced Zarqawi's influence. Col. Derek Harvey, who served as a military intelligence officer in Iraq and then was one of the top officers handling Iraq intelligence issues on the staff of the Joint Chiefs of Staff, warned an Army meeting in 2004 that "Our own focus on Zarqawi has enlarged his caricature, if you will — made him more important than he really is, in some ways." While Pentagon spokespersons state unequivocally that PSYOPs may not be used to influence American citizens, there is little question that the information disseminated through the program has found its way into American media sources. The *Washington Post* also notes that "One briefing slide about U.S. "strategic communications" in Iraq, prepared for Army Gen. George W. Casey Jr., the top U.S. commander in Iraq, describes the "home audience" as one of six major targets of the American side of the war."

On July 4, 2006, the US Ambassador to Baghdad Zalmay Khalilzad, in an interview with the *BBC*, said that "in terms of the level of violence, it (the death of al-Zarqawi) has not had any impact at this point" and that "...the level of violence is still quite high." But Khalilzad maintained his view that the killing had though encouraged some insurgent groups to "reach out" and join government reconcialiation talks, he believed that previously these groups were intimidated by Zarqawi's presence.

On 8th of June 2006, on the BBC's Question Time, the Respect Party MP George Galloway referred to Zarqawi as 'a Boogeyman, built up by the Americans to try and perpetrate the lie that the resistance in Iraq are by foreigners, and that the mass of the Iraqi's are with the American and British occupation'.

On August 21, 2006, Jill Carroll, a journalist for the Christian Science Monitor, published part 6 of her story detailing her capitivity in Iraq. In it, she describes how one of her captors, who identified himself as Abdullah Rashid and leader of the Mujahideen Shura Council in Iraq, conveyed to her that "The Americans were constantly saying that the mujahideen in Iraq were led by foreigners...So, the Iraqi insurgents went to Zarqawi and insisted that an Iraqi be put in charge." She continued by stating: "But as I saw in coming weeks, Zarqawi remained the insurgents' hero, and the most influential member of their council, whatever Nour/ Rashid's position. And it seemed to me, based on snatches of conversations, that two cell leaders under him-Abu Rasha and Abu Ahmed-might also be on the council. At various times, I heard my captors discussing changes in their plans because of directives from the council and Zarqawi."

Pre-war Opportunities to kill Zarqawi

According to *NBC News*, the Pentagon had pushed to "take out" Zarqawi's operation at least three times prior to the invasion of Iraq, but had been vetoed by the National Security Council. The council reportedly made its decision in an effort to convince other countries to join the US in a coalition against Iraq. "People were more obsessed with developing the coalition to overthrow Saddam than to execute the president's policy of pre-emption against terrorists," said former National Security Council member Roger Cressey.

In May 2006, former CIA official Michael Scheuer, who headed the CIA's bin Laden unit for six years before resigning in 2004, corroborated this. Paraphrasing his remarks, the *Australian Broadcasting Corporation* stated Scheuer claimed that "the United States deliberately turned down several opportunities to kill terrorist Abu Musab al-Zarqawi in the lead-up to the Iraq war." ABC added that "a plan to destroy Zarqawi's training camp in Kurdistan was abandoned for diplomatic reasons."

Scheuer explained that "the reasons the intelligence service got for not shooting Zarqawi was simply that the President and the National Security Council decided it was more important not to give the Europeans the impression we were gunslingers" in an effort to win support for ousting Saddam Hussein.

This claim was also corroborated by CENTCOM's Deputy Commander, Lt. General Michael DeLong, in an interview with *PBS* on February 14, 2006.

DeLong, however, claims that the reasons for abandoning the opportunity to take out Zarqawi's camp was that the Pentagon feared that an attack would contaminate the area with chemical weapon materials: "We almost took them out three months before the Iraq war started. We almost took that thing, but we were so concerned that the chemical cloud from there could devastate the region that we chose to take them by land rather than by smart weapons."

Reports of Zarqawi's Death, Detention and Injuries

Missing Leg

Claims of harm to Zarqawi changed over time. Early in 2002, there were unverified reports from Afghan Northern Alliance members that Zarqawi had been killed by a missile attack in Afghanistan. Many news sources repeated the claim. Later, Kurdish groups claimed that Zarqawi had not died in the missile strike, but had been severely injured, and went to Baghdad in 2002 to have his leg amputated. On October 7, 2002, the day before Congress voted to give President Bush authorization to invade Iraq, Bush gave a speech in Cincinnati, Ohio, that repeated as fact the claim that he had sought medical treatment in Baghdad. This was one of several of President Bush's primary examples of ways Saddam Hussein had aided, funded, and harbored al-Qaeda. Powell repeated this claim in his February 2003 speech to the UN, urging a resolution for war, and it soon became "common knowledge" that Zarqawi had a prosthetic leg.

In 2004, *Newsweek* reported that some "senior U.S. military officials in Baghdad" had come to believe that he still had his original legs. Knight Ridder later reported that the leg amputation was something "officials now acknowledge was incorrect."

When the video of the Berg beheading was released in 2004, credence was given to the claim that Zarqawi was alive and active. The man identified as Zarqawi in the video did not appear to have a prosthetic leg. Videos of Zarqawi aired in 2006 that clearly showed him with both legs intact. When Zarqawi's body was autopsied, "X-rays also showed a fracture of his right lower leg."

Claims of Death

In March 2004, an insurgent group in Iraq issued a statement saying that Zarqawi had been killed in April 2003. The statement said that he was unable to escape the missile attack because of his prosthetic leg. His followers claimed he was killed in a US bombing raid in the north of Iraq. The claim that Zarqawi had been killed in northern Iraq "at the beginning of the war," and that subsequent use of his name was a useful myth, was repeated in September

2005 by Sheikh Jawad Al-Khalessi, a Shiite imam. On May 24, 2005, it was reported on an Islamic website that a deputy would take command of Al-Qaeda while Zarqawi recovered from injuries sustained in an attack. Later that week the Iraqi government confirmed that Zarqawi had been wounded by U.S. forces, although the battalion did not realize it at the time. The extent of his injuries is not known, although some radical Islamic websites called for prayers for his health. There are reports that a local hospital treated a man, suspected to be Zarqawi, with severe injuries.

He was also said to have subsequently left Iraq for a neighbouring country, accompanied by two physicians. However, later that week the radical Islamic website retracted its report about his injuries and claimed that he was in fine health and was running the jihad operation.

In a September 16, 2005 article published by *Le Monde*, Sheikh Jawad Al-Kalesi claimed that al-Zarqawi was killed in the Kurdish northern region of Iraq at the beginning of the US-led war on the country as he was meeting with members of the Ansar al-Islam group affiliated to al-Qaeda. Al-Kalesi also claimed "His family in Jordan even held a ceremony after his death." He also claimed that "Zarqawi has been used as a ploy by the United States, as an excuse to continue the occupation" and saying, "It was a pretext so they don't leave Iraq."

On November 20, 2005, some news sources reported that Zarqawi may have been killed in a coalition assault on a house in Mosul; five of those in the house were killed in the assault while the other three died through using 'suicide belts' of explosives. United States and British soldiers searched the remains, with U.S. forces using DNA samples to identify the dead. However, none of those remains belonged to him.

Reportedly Captured and Released

According to a *CNN* report dated December 15, 2005, al-Zarqawi was captured by Iraqi forces sometime during 2004 and later released because his captors did not realize who he was. This claim was made by a Saudi suicide bomber, Ahmed Abdullah al-Shaiyah, who survived a failed suicide attempt to blow up the

Jordanian mission in Baghdad in December. "Do you know what has happened to Zarqawi and where he is?" an Iraqi investigator asked Mr. Shaiyah. He answered, "I don't know, but I heard from some of my mujahadeen brothers that Iraqi police had captured Zarqawi in Fallujah." Mr. Shaiyah says he then heard that the police let the terrorist go because they had failed to recognize him. U.S. officials called the report "plausible" but refused to confirm it.

Zarqawi's Death

Zarqawi was killed on June 7, 2006, while attending a meeting in an isolated safehouse approximately 8 km (5 mi) north of Baqubah. At 14:15 GMT two United States Air Force F-16C jets identified the house and the lead jet dropped two 500-pound (230kg) guided bombs, a laser-guided GBU-12 and GPS-guided GBU-38 on the building located at 33°482 02.833 N 44°302 48.583 Eÿþ / ÿþ33.8007861, 44.5134944. Six others-three male and three female individuals-were also reported killed (see below). Among those killed were his wife and their child.

The joint task force had been tracking him for some time, and although there were some close calls, he had eluded them on many occasions. United States intelligence officials then received tips from Iraqi senior leaders from Zarqawi's network that he and some of his associates were in the Baqubah area. The safehouse itself was watched for over six weeks before Zarqawi was observed entering the building by U.S. AFSOC 720th Special Tactics Group Combat Controllers. Jordanian intelligence reportedly helped to identify his location. The area was subsequently secured by Iraqi security forces, who were the first ground forces to arrive.

On June 8, 2006, coalition forces confirmed that Zarqawi's body was identified by facial recognition, fingerprinting, known scars and tattoos. They also announced the death of one of his key lieutenants, spiritual adviser Sheik Abd-Al-Rahman.

Initially, the U.S. military reported that Zarqawi was killed directly in the attack. However, according to a statement made the following day by Major General William Caldwell of the U.S. Army, Zarqawi survived for a short time after the bombing, and after being placed on a stretcher, attempted to move and was restrained, after which he died from his injuries.

An Iraqi man, who claims to have arrived on the scene a few moments after the attack, said he saw U.S. troops beating up the badly-wounded but still alive Zarqawi. In contradiction, Caldwell asserted that when U.S. troops found Zarqawi barely alive they tried to provide him with medical help, rejecting the allegations that he was beaten based on an autopsy performed. The account of the Iraqi witness has not been verified. All others in the house died immediately in the blasts. On June 12, 2006, it was reported that an autopsy performed by the U.S. military revealed that the cause of death to Zarqawi was a blast injury to the lungs, but he took nearly an hour to die.

The U.S. government distributed an image of Zarqawi's corpse as part of the press pack associated with the press conference. The release of the image has been criticised for being in questionable taste, and for inadvertently creating an iconic image of Zarqawi that would be used to rally his supporters.

Reactions to Death

Prime Minister of Iraq Nuri al-Maliki commented on the death of Zarqawi by saying: "Today, Zarqawi has been terminated. Every time a Zarqawi appears we will kill him. We will continue confronting whoever follows his path. It is an open war between us."

United States President George W. Bush stated that through his every action al-Zarqawi sought to defeat America and its coalition partners by turning Iraq into a safe haven for al-Qaeda. Bush also stated, "Now Zarqawi has met his end and this violent man will never murder again."

Zarqawi's brother-in-law has since claimed that he was a martyr even though the family renounced Zarqawi and his actions in the aftermath of the Amman triple suicide bombing that killed at least 60 people. The opinion of Iraqis on his death is mixed; some believe that it will promote peace between the warring factions, while others are convinced that his death will provoke his followers to a massive retaliation and cause more bombings and deaths in

Iraq. Abu Abdulrahman al-Iraqi, the deputy of al-Zarqawi (which may be the individual called "Sheik Abd-Al-Rahman" mentioned above, meaning he was not present as the bombing happened), released a statement to Islamist websites indicating that al-Qaeda in Iraq also confirmed Zarqawi's death: "We herald the martyrdom of our mujahed Sheikh Abu Musab al-Zarqawi in Iraq ... and we stress that this is an honor to our nation." In the statement, al-Iraqi vowed to continue the jihad in Iraq.

On June 16, 2006, Abu Abdullah Rashid al-Baghdadi, the head of the Mujahideen Shura Council, which groups five Iraqi insurgent organizations including al-Qaeda in Iraq, released an audio tape statement in which he described the death of al-Zarqawi as a "great loss." He continued by stating that al-Zarqawi "will remain a symbol for all the mujahideen, who will take strength from his steadfastness." Al-Baghdadi is believed to be a former officer in Saddam's army, or its elite Republican Guard, who has worked closely with al-Zarqawi since the overthrow of Saddam's regime in April 2003.

Counterterrorism officials have said that al-Zarqawi had become a key part of al-Qaeda's marketing campaign and that al-Zarqawi served as a "worldwide jihadist rallying point and a fundraising icon." Rep. Mike J. Rogers, R-Mich., who serves on the House Intelligence Committee, called al-Zarqawi "The terrorist celeb, if you will,... It is like selling for any organization. They are selling the success of Zarqawi in eluding capture in Iraq."

On June 23, 2006, Al-Jazeera aired a video in which Ayman al-Zawahiri, al-Qaeda's No. 2 leader, states that Zarqawi was "a soldier, a hero, an imam and the prince of martyrs, [and his death] has defined the struggle between the crusaders and Islam in Iraq."

On June 30, 2006, Osama bin Laden released an audio recording in which he stated, "Our Islamic nation was surprised to find its knight, the lion of jihad, the man of determination and will, Abu Musab al-Zarqawi, killed in a shameful American raid. We pray to God to bless him and accept him among the martyrs as he had hoped for." Bin Laden also defended al-Zarqawi, saying he had "clear instructions" to focus on U.S.-led forces in Iraq but also "for those who... stood to fight on the side of the crusaders against the

Muslims, then he should kill them whoever they are, regardless of their sect or tribe." Shortly after, he released another audio tape in which he stated, "Our brothers, the mujahedeen in the al-Qaeda organization, have chosen the dear brother Abu Hamza al-Muhajer as their leader to succeed the Amir Abu Musab al-Zarqawi. I advise him to focus his fighting on the Americans and everyone who supports them and allies himself with them in their war on the people of Islam and Iraq."

Alleged Betrayal by al-Qaeda

A day before Zarqawi was killed, a U.S. strategic analysis site suggested that Zarqawi could have lost the trust of al-Qaeda due to his emphatic anti-Shia stance and the massacres of civilians allegedly committed in his name. Reports in *The New York Times* on June 9 treated the betrayal by at least one fellow al-Qaeda member as fact, stating that an individual close to Zarqawi disclosed the identity and location of Sheik Abd al-Rahman to Jordanian and American intelligence. Non-stop surveillance of Abd al-Rahman quickly led to Zarqawi.

The *Associated Press* quotes an unnamed Jordanian official as saying that the effort to find Zarqawi was successful partly due to information that Jordan obtained one month beforehand from a captured Zarqawi al-Qaeda operative named Ziad Khalaf Raja al-Karbouly.

Reward

In apparent contradiction to statements made earlier in the day by U.S. ambassador to Iraq Zalmay Khalilzad, an Iraqi spokesman said the US$25 million reward "will be honored" (although this need not mean that any money will actually be paid, as the terms of the reward would indeed be "honored" by having no payee if no one qualifies). Khalilzad, in an interview with CNN's Wolf Blitzer, had stated the bounty would not be paid because the decisive information leading to Zarqawi's whereabouts had been supplied by an al-Qaeda in Iraq operative whose own complicity in violent acts would disqualify him from receiving payment.

Rep. Mark Kirk, a Republican of Illinois who wrote the legislation specifying the Zarqawi reward, has been quoted as saying that the Bush Administration does plan to pay "some rewards" for Zarqawi. "I don't have the specifics," he said, "The administration is now working out who will get it and how much. As their appropriator who funds them, I asked them to let me know if they need more money to run the rewards program now that they are paying this out."

Post-Zarqawi Iraq Environment

Zarqawi's death was seen a major coup for the US government in terms of the political and propaganda stakes. However, unconfirmed rumors in early April 2006 suggested that Zarqawi had been demoted from a strategic or coordinating function to overseer of paramilitary/terrorist activities of his group and that Abdullah bin Rashed al-Baghdadi of the Mujahideen Shura Council succeeded Zarqawi in the former function. On June 15, 2006, the United States military officially identified Abu Ayyub al-Masri as the successor to Al-Qaeda in Iraq.

After Zarqawi's demise in early June 2006 there has been little or no immediately identifiable change in terms of the level of violence and attacks against U.S. and allied troops. In the immediate aftermath insurgency attacks averaged 90 a day, apparently some of the highest on record. Four months after Zarqawi's death, it is estimated that 374 coalition soldiers and 10355 Iraqis have been killed. Several insurgency groups and heads of Sunni Muslim tribes also formed a coalition called the Mujahideen Shura Council.

By late 2007, violent and indiscriminate attacks directed by AQI against Iraqi civilians had severely damaged their image and caused the loss of support among the population, isolating the group. In a major blow to AQI, thousands of former Sunni militants that previously fought along with the group started to actively fight AQI and also work with the American and Iraqi forces starting with the creation of the Anbar Awakening Council because of its Anbar origins. The group spread to all Sunni cities and communities and some Shite areas and adopted the broader name Sons of Iraq. The Sons of Iraq was instrumental in giving tips to coalition forces about weapons caches and militants resulting in the the destruction of over 2,500 weapons caches and over 800 militants being killed or captured. In addition, the 30,000 strong U.S. troop surge supplied military planners with more manpower for operations targeting Al-Qaeda in Iraq, The Mujahadeen Shura Council, Ansar Al-Sunnah and other terrorist groups. The resulting events leading to dozens of high-level AQI leaders being captured or killed. Al-Qaeda seemed to have lost its foothold in Iraq and appeared to be severely crippled due to its lack of vast weapons caches, leaders, safe havens, and Iraqis willing to support them. Accordingly, the bounty issued for Abu Ayyub-al-Masri AKA Abu Hamza al-Muhajer was eventually cut from $5 million down to a mere $100,000 in April 2008.

On January 8, 2008, & January 28, 2008, Iraqi and U.S. forces launched operations Phantom Phoenix and the Ninawa campaign AKA the Mosul Campaign killing and capturing over 4,600 militants and locating and destroying over 3,000 weapons caches in those 2 campaigns. Also effectively leaving AQI with 1 last major insurgent stronghold Diyala. On July 29, 2008 Iraqi, U.S. and Sons Of Iraq forces launched Operation Augurs of Prosperity in the Diyala province and surrounding areas to clear AQI out of its last stronghold. 2 operations were already launched before in Diyala with mixed results and this campaign was expected to face fierce resistance. The rustling operation left over 500 weapons caches destroyed and 5 militants killed, 483 militants were captured due to the lack of resistance from the insurgent forces. 24 high level AQI terrorist were killed or captured in the campaign. As of August 1st 2008 there has been 83.2% decline in violence in Iraq.

AL-QAEDA AND RELIGIOUS TERRORISM

Religious terrorism is terrorism by those whose motivations and aims have a predominant religious character or influence.

According to Mark Juergensmeyer, religious terrorism consists of acts that terrify, the definition of which is provided by the witnesses-the ones terrified-and not by the party committing the act; accompanied by either a religious motivation, justification, organization, or world view. Religion is sometimes used in combination with other factors, and sometimes as the primary motivation. Religious terrorism is intimately connected to current forces of geopolitics.

According to Bruce Hoffman, to be considered religious terrorism the perpetrators must use religious scriptures to justify or explain their violent acts or to gain recruits and there must be some sort of clerical figures involved in some leadership roles.

In the modern age, after the decline of ideas such as the divine right of kings and with the rise of nationalism, terrorism more often involved anarchism, nihilism and revolutionary politics, but since 1980 there has been an increase in activity motivated by religion.

Former United States Secretary of State Warren Christopher said that terrorist acts in the name of *religion and ethnic identity* have become "one of the most important security challenges we face in the wake of the Cold War." Stephen Weinberg has argued that religion is the most important factor, famously saying "for good people to do evil things, that takes religion". However, Robert Pape, Rogers et al, Nardin, and Juergensmeyer have all argued that religion should be considered only one incidental factor, and that so-called "religious" terrorism is primarily geopolitical.

Criticism of the Concept of "Religious Terrorism"

Robert Pape, political science professor of the University of Chicago, compiled the first complete database of every documented suicide bombing from 1980-2003. He argues that the news reports about suicide attacks are profoundly misleading-"There is little connection between suicide terrorism and Islamic fundamentalism, or any one of the world's religions". After studying 315 suicide attacks carried out over the last two decades, he concludes that suicide bombers' actions stem from political conflict, not religion.

Michael A. Sheehan stated in 2000, "A number of terrorist groups have portrayed their causes in religious and cultural terms. This is often a transparent tactic designed to conceal political goals, generate popular support and silence opposition." Terry Nardin wrote, "A basic problem is whether religious terrorism really differs, in its character and causes, from political terrorism... defenders of religious terrorism typically reason by applying commonly acknowledged moral principles... But the use (or misuse) of moral arguments does not in fact distinguish religious from nonreligious terrorists, for the latter also rely upon such arguments to justify their acts... political terrorism can also be symbolic... alienation and dispossession... are important in other kinds of violence as well. In short, one wonders whether the expression 'religious terrorism' is more than a journalistic convenience".

Professor Mark Juergensmeyer wrote, "..religion is not innocent. But it does not ordinarily lead to violence. That happens only with the coalescence of a peculiar set of circumstances-political, social, and ideological-when religion becomes fused with violent expressions of social aspirations, personal pride, and movements for political change." and "Whether or not one uses 'terrorist' to describe violent acts depends on whether one thinks that the acts are warranted. To a large extent the use of the term depends on one's world view: if the world is perceived as peaceful, violent acts appear to be terrorism. If the world is thought to be at war, violent acts may be regarded as legitimate. They may be seen as preemptive strikes, as defensive tactics in an ongoing battles, or as symbols indicating to the world that it is indeed in a state of grave and ultimate conflict".

Financing

Terrorism activities worldwide are supported through not only the organized systems that teach holy war as the highest calling, but also through the legal, illegal, and often indirect methods financing these systems which sometimes utilize organizations as fronts to mobilize or channel sources and funds, including charities. Charities can involve the provision of aid to those in need and oblations or charitable offerings occur in nearly all religious systems with sacrifice as a furtherance of the custom.

Martyrdom and Suicide Terrorism

Important symbolic acts such as the blood sacrifice link acts of violence to religion and terrorism. Suicide terrorism, self-sacrifice, or martyrdom has throughout history been organized and perpetrated by groups with both political and religious motivations. Suicide terrorism or martyrdom is efficient, inexpensive, easily organized, and extremely difficult to counter, delivering maximum damage for little cost. The shocking nature of a suicide attack also attracts public attention. Glorifying the culture of martyrdom benefits the terrorist organization and inspires more people to join the group. Retaliation against suicide attacks, often in the form of collective punishment, further increases the group's sense of victimization and commitment to adhere to doctrine and policy. This process serves to encourage martyrdom. Suicide terrorism, self-sacrifice, or martyrdom therefore represent "value for money".

DESTRUCTIVE CULT

"Destructive cult" is a term used to refer to religions and other groups which have caused harm to their own members or to others. Some researchers define "harm" in this case with a narrow focus, specifically groups which have deliberately physically injured or killed other individuals, while others define the term more broadly and include emotional abuse among the types of harm inflicted. Use of the term has been criticized by some researchers, who assert that it is used to discredit the groups it is applied to, and unfairly compare them with historically more harmful groups and movements. Authors have also compared destructive cults with terrorism, and have used the term to characterize Osama bin Laden as a destructive cult leader.

Physical Harm

"Destructive cult" as applied to physical abuse has generally referred to groups which have, through deliberate action, physically injured or killed members of their own group or other individuals. The Ontario Consultants on Religious Tolerance limit use of the term to specifically refer to religious groups that "have caused or are liable to cause loss of life among their membership or the general public."

Emotional Harm

Some researchers describe the term "destructive cult" more

broadly, and include emotional abuse along with physical abuse as a defining characteristic. Steven Hassan, author of the book *Combatting Cult Mind Control,* defines the term as such: "A destructive cult is a pyramid-shaped authoritarian regime with a person or group of people that have dictatorial control. It uses deception in recruiting new members (e.g. people are NOT told up front what the group is, what the group actually believes and what will be expected of them if they become members)." Psychologist Michael Langone, executive director of the International Cultic Studies Association, defines a destructive cult as "a highly manipulative group which exploits and sometimes physically and/or psychologically damages members and recruits." In the book *Into the Rabbit Hole* contributor Randall Waters cites psychiatrist Robert Lifton's *Thought Reform and the Psychology of Totalism,* specifically Lifton's "Eight Criteria for Thought Reform", as criteria to identify a destructive cult. In *Perfected Mind Control-The Unauthorized Black Book of Hypnotic Mind Control* author J. K. Ellis also cites Lifton's criteria, writing: "If most of Robert Lifton's eight point model of thought reform is being used in a cult organization, it is most likely a dangerous and destructive cult." In a statement which Congressman Leo J. Ryan later read into the *Congressional Record,* Dr. John Gordon Clark cited totalitarian systems of governance and an emphasis on money making as characteristics of a destructive cult. Dr. Clark later authored the work *Destructive Cult Conversion: Theory, Research and Practice.*

In their book *Cult and Ritual Abuse,* authors Noblitt and Perskin also include the characteristics of "trauma-induced dissociation and programming" in their definition of a destructive cult. In *Cults and the Family* the authors cite Shapiro, who defines a "destructive cultism" as a sociopathic syndrome, whose distinctive qualities include: "behavioral and personality changes, loss of personal identity, cessation of scholastic activities, estrangement from family, disinterest in society and pronounced mental control and enslavement by cult leaders." In *The Ethics of Touch: The Hands-on Practitioner's Guide To Creating a Professional, Safe and Enduring Practice,* the authors describe their version of destructive cult characteristics in a section on "Cult Mind Control Abuse." In the book, a destructive cult is seen as being either "religious, political, 'therapeutic' or business" and they state that it can cause trauma-related symptoms such as dissociative disorder. In Dr. Susan Gregg's *The Complete Idiot's Guide to Spiritual Healing*, she cites three main signs of a destructive cult, including giving up one's individuality, having their relationships with friends and family threatened, and being asked to donate large sums of money to the group. In his work *Lethal Violence*, the criteria of a destructive cult environment is compared to that of battered woman defence.

Criticism of the Term

Some researchers have criticized the usage of the term "destructive cult", writing that it is used to describe groups which are not necessarily harmful in nature to themselves or others. In his book *Understanding New Religious Movements*, John A. Saliba writes that the term is overgeneralized and equated with the deaths of members of Peoples Temple at Jonestown. Saliba sees this as the "paradigm of a destructive cult," where those that use the term are implying that other new religious movements will have similar outcomes to those of the Peoples Temple at Jonestown. Writing in the book *Misunderstanding Cults: Searching for Objectivity in a Controversial Field*, contributor Julius H. Rubin complains that the term has been used to discredit certain groups in the court of public opinion. In his work *Cults in Context* author Lorne L. Dawson writes that though the Unification Church "has not been shown to be violent or volatile," it has been described as a destructive cult by "anticult crusaders." In 2002, the German government was held by Germany's Federal Constitutional Court to have defamed the Osho movement by referring to it, among other things, as a "destructive cult". The court decided that "destructive cult" and other expressions employed by the government to describe the group had no factual basis to justify their use.

Bibliography

Ahmad, R.S. : *Al-Harakat Al-Islamiyya.* Cairo: Sina Linasr. Allport, G. 1958.

Alam, Aftab : *US Policy Towards South Asia: Special Reference to Indo-Pak Relations,* Raj, Delhi, 1998.

Allana, G.: *Pakistan Movement: Historic Documents,* Karachi, Department of International Relations, University of Karachi, 1967.

Ayesha Siddiqa-Agha: *Military Inc.: Inside Pakistan's Military Economy,* London: Pluto, 2007.

Ayman Talal Yousef and Dilip H. Mohite : *Rise of Islamic Fundamentalism and the Grand American Strategy* : Kalinga, Delhi, 2002.

Barbara D. Metcalf and Thomas R. Metcalf : *A Concise History of India and Pakistan,* New York: Cambridge University Press, 2002).

Burki, Shahid Javed: *Pakistan: The Continuing Search for Nationhood,* Westview Press, 1991.

Chopra, Surendra and Kusum Lata Chadda: *Islamic Fundamentalism, Pakistan and the Muslim World* : Kanishka, Delhi, 2009.

Choudhury, Golam W: *Pakistan: Transition from Military to Civilian Rule,* Scorpion, 1988.

Daniel Pipes : *In the Path of God: Islam and Political Power,* New York: Basic Books Inc., 1983.

Dekmejian, R. Hrair : *Islamic Revolution: Fundamentalism in the Arab World* Syracuse, New York: Syracuse University Press, 1995.

Dilip H. Mohite : *Rise of Islamic Fundamentalism and the Grand American Strategy* : Kalinga, Delhi, 2002.

Friedman, G.: *The Future of War: Power, Technology and Pakistan Army,* N.Y., Random House, 1996.

Gupta, J.B. Das : *Islamic Fundamentalism and India* : Hope India Publications, Delhi, 2002.

Hasan Askari Rizvi: *Military, State, and Society in Pakistan,* New York: St. Martin's Press, 2000.

Hassan, Lt Gen Gul, *Memoirs of Lt Gen Gul Hassan*, Oxford University Press, 1993.

Hiro, Dilip : *Holy Wars: The Rise of Islamic Fundamentalism* New York: Routledge, Chapman and Hall, Inc., 1989.

Hussain Mushahid: *Pakistan: Problems of Governance*, Vanguard 1993.

Ian Talbot: *Pakistan: A Modern History*, London: Hurst, 2009.

John O. Voll : *Islam: Continuity and Change in the Modern World*, Syracuse, New York: Syracuse University Press, 1994.

Khan, Major General Sher Ali: *The Story of Soldiering and Politics in India and Pakistan*, Wajidalis, 1978.

Martin Marty and R. Scott Appleby: *The Fundamentalism Project*, Chicago and London: University of Chicago Press, 1993,

Multani, R K P : *Islamic Fundamentalism in South Asia* : Sumit Enterprises, Delhi, 2007.

Nair, A. M.: *Role of Indian Pakistan in Kargir War*, Bombay, Orient Longman, 1983.

Rizvi, Hassan Askari: *The Military and Politics in Pakistan: 1947 - 86*, Progressive Publishers 1986.

Roxanne L. Euben, *Enemy in the Mirror : Islamic Fundamentalism and the Limits of Modern Rationalism : A Work of Comparative Political Theory* : Oxford University Press, 2001.

Saeed Shafqat, *Civil-Military Relations in Pakistan: From Zulfikar Ali Bhutto to Benazir Bhutto*, Boulder: Westview, 1997.

Shuja Nawaz: *Crossed Swords: Pakistan, Its Army, and the Wars Within*, Karachi: Oxford University Press, 2008.

Siddiqi, Birgadier A. R.: *The Military in Pakistan: Image and Reality*, Vanguard, 1996.

Stephen P. Cohen: *The Idea of Pakistan*, Washington, DC: Brookings Institution Press, 2004.

Sumit Ganguly: *Conflict Unending: India-Pakistan Tensions since 1947*, New York: Columbia University Press, 2001.

William R. Thompson: *The Grievances of Military Coup-makers*, Beverly Hills: Sage Publications, 1973.

Zaidi, Z. H.: *M. A. Jinnah-Ispahani Correspondence*, 1936-1948, Karachi, Forward publishing Trust, 1976.

Index

❑❑❑